OPEN AIR GRAPE CULTURE:

A PRACTICAL TREATISE ON THE

Garden and Vineyard Culture of the Vine,

AND THE

MANUFACTURE OF DOMESTIC WINE.

DESIGNED FOR THE USE OF AMATEURS AND OTHERS
IN THE NORTHERN AND MIDDLE STATES.

PROFUSELY ILLUSTRATED WITH NEW ENGRAVINGS FROM CAREFULLY EXECUTED DESIGNS, VERIFIED BY DIRECT PRACTICE.

BY JOHN PHIN,

AUTHOR OF "ESSAY ON OPEN AIR GRAPE CULTURE," TO WHICH WAS AWARDED THE FIRST PREMIUM OF THE AMERICAN INSTITUTE.

TO WHICH IS ADDED A SELECTION OF

EXAMPLES OF AMERICAN VINEYARD PRACTICE,

AND A CAREFULLY PREPARED DESCRIPTION OF THE CELEBRATED

THOMERY SYSTEM OF GRAPE CULTURE.

NEW YORK:
C. M. SAXTON, AGRICULTURAL BOOK PUBLISHER.
1862.

W. H. Tinson, Stereotyper.

PREFACE.

THE following work has been undertaken, not so much in the hope of adding anything new to what is already known of the culture of the vine, as with a view to collect the scattered information which exists on the subject in periodicals and kindred works as well as amongst practical men, and to throw it into such a shape as may prove useful to the amateur and the vine-dresser.

This being our object, we have endeavored to modify and adapt the practice and principles of others to our own climate and wants, and to simplify and explain the processes of the professional gardener so that he who reads may practice. To this end we have in general avoided theoretical discussions, and have depended chiefly upon the practice of ourselves and others for the directions here laid down. For although we know that well established principles are the only sure foundation of all right practice, this is not the place for discussing the theoretical grounds upon which these principles rest. A practical work should deal with facts and be a guide to action.

As the garden culture of the vine, at least in the northern States, differs from that in the vineyard only in the more thorough preparation of the ground and the larger size of the plants, we have not formally divided the work into sections corresponding to these two classes, as the principles which govern both are precisely alike.

Where, however, some peculiar details of management apply to either we have inserted them in the section to which they properly belong—as under the subject of VINE BORDERS and

CARE OF OLD VINES. A full account of the Ohio vineyards is given in the Appendix, amongst other examples of American practice, and the peculiar principles which regulate the management of grapes devoted to the production of wine will be found in their appropriate place, viz., in the second part of this work, which is specially devoted to that subject.

The varieties of the vine have multiplied so rapidly of late, that it would be impossible to give a complete list even of those which have been brought out. Seeing then that at best our work must be incomplete in this respect, we have described those only which have been thoroughly proved and recommended by some well known society or cultivator. Of the two or three hundred varieties of American grapes of which names are to be found, probably not more than one in ten have been tested in localities differing greatly from the place of their origin.

In the execution of our work, we believe that where we have had occasion to make use of the labors of others, due credit has always been given; and we have also added a list of those books which we have most freely consulted, so that those who desire to make the culture of the grape a specialty may be directed to original sources of information.

That the culture of the grape will ere long attain a position of which its present condition affords little idea, we have no doubt. Not only is it one of the most delicious and easily raised fruits, but it also gives quick returns, so that he who plants a vine has not to wait for the better portion of his lifetime ere he eats the fruit of it; in three or four years it will yield an ample vintage.

CONTENTS.

THE MANUFACTURE OF WINE.

APPENDIX I.

APPENDIX II.

THE THOMERY SYSTEM OF GRAPE CULTURE.

(This is the famous system under which the splendid Chasselas de Fontainebleau Grapes are produced; it is the method advocated by Dr. Grant. The account is literally translated from Du Breuil's Cours d'Arboriculture, and is the only complete account which can be found—at least so says the Frenchman, Rendu, in his "Ampelographie Française," and he ought to know.)

PAGE

BIBLIOGRAPHY OF GRAPE CULTURE.

THE subjoined list contains all the principal works which have been consulted in the preparation of the following treatise. Having made the study of the subject a specialty, we have been at considerable pains to collect all the works relating to vine culture of which we could find any account, and although there are several important omissions in the list given below, yet it is believed that their place is tolerably well supplied by those of which titles are given. As our attention has been chiefly directed to open air culture, we have intentionally omitted some excellent English works. The French, however, possess some valuable treatises which we regret having been unable to obtain, and a still greater source of regret has been that we have been unable to use the many fine works possessed by the Germans.

It was suggested that this list be prefaced with a short article on the bibliography of grape culture, or at least that the peculiar features of the works mentioned be indicated. But we found ourselves incompetent to the former, and the latter would have occupied a space disproportionate to its importance in a practical treatise. It is hoped, however, that the list given will not prove useless to those who desire to extend their inquiries beyond the narrow limits of the present work, and from the assistance which we ourselves have frequently derived from similar catalogues, we feel confident that this hope is not ill founded. It may be added, in conclusion, that many works have been consulted and used of which no mention is made, simply from the remoteness of their general bearing upon the subject. Thus the figure of the oïdium is taken from Pouillet's "Traité de

Physique," and is, we believe, the only thing in all the three volumes of that work which at all relates to vine culture.

Having no desire to preface our work with a mere catalogue of our private library (as we have seen done more than once), no work has been mentioned which it will not repay the reader to consult. On the general subject of the "Theory of Horticulture," Lindley's work has been our guide and our standard, and for our chemical facts and principles we have relied upon the work of Gmelin, published by the Cavendish Society of London, in twelve volumes, as we have always found it most full and reliable.

But in selecting a course of reading with a view to advance his knowledge of grape culture, the student must bear in mind that so varied, complex and intimately connected are all the operations of nature, that the facts which have a bearing upon any portion of them, are to be found in books which professedly treat of the most diverse subjects. Chemistry and mechanics are alike important; the principles which govern the relations of heat, light, and electricity, exert a more or less important influence on all vegetation, and he who would be fully master of the subject, must aim at an extent of knowledge only to be found in the widest range of scientific reading and experiment.

Abercrombie, John, Practical Gardener. London.
Adlum, John, Memoir on the Cultivation of the Vine in America. Washington, 1828.
Allen, J. F., Practical Treatise on the Culture of the Grape. New York, 1858.
——— the same, Boston, 1849.
American Cyclopædia. New York, 1858, *continued.*
American Pomological Society, Transactions 1852, 1854, 1856, 1858.

Barry, Sir Ed., Observations on Wines. 4to. London, 1775.
Barry, P., Fruit Garden. New York, 1855.
Bernay, A. J., Household Chemistry. London, 1854.
Blodgett, Lorin, Climatology of the United States. Philadelphia, 1857.
Bordeaux Wine and Liquor Dealer's Guide. New York, 1851.
Boussingault, J. B., Rural Economy. London, 1855.
Bradley, R., Survey of Ancient Husbandry and Gardening. London, 1725.

Bridgeman, Thomas, Young Gardener's Assistant. New York, 1857.
Brown, J. D., Sylva Americana. Boston, 1832.
——— Trees of America. New York, 1851.
——— Field Book of Manures. New York, 1855.
Buchannan, R., Culture of the Grape and Wine Making. Cincinnati, 185–.
Buist, R., Management of the Grape Vine. New York, 1856.
Busby, James, Visit to Vineyards of France and Spain. New York, 1835.

Carnell, P. P., Treatise on Family Wine Making. London, 1814.
Carpenter, W. B., Use and Abuse of Alcoholic Liquors. Philadelphia, 1855.
Chaptal, C., Traité sur la Vigne et l'Art de faire Vin. 2 vols., Paris, 1801.
——— The same, translated in Philosophical Magazine.
——— Chemistry applied to Agriculture. Hartford, 1854.
Chorlton, Wm., American Grape Grower's Guide. New York, 1856.
——— The Cold Grapery. New York, 1853.
Cole, S. W., American Fruit Book. Boston, 1849.
Coleman, Henry, European Agriculture. Boston.
Country Gentleman. Albany (*published weekly*).

Dana, S. L., Muck Manual. New York, 1856.
Davy, Sir H., Agricultural Chemistry. London, 1827.
Davy, John, Ionian Islands. London, 1842.
De Bow, J. B. D., Industrial Resources of South and West. New Orleans, 1852.
——— Review. New Orleans.
Décandolle, N. P., Physiologie Végétale. 3 vols. Paris, 1832.
Dempsey, G. D., On the Drainage of Districts and Lands. London, 1854.
Don, George, General System of Gardening and Botany. 4 vols., 4to., London, 1838.
Donaldson, Treatise on Clay Lands and Loamy Soils. London, 1854.
Donovan, Michael, Treatise on Domestic Economy and Wine Making. London, 1830.
Downing, A. J., Fruits and Fruit Trees of America. New York, 1853.
——— the same, revised by C. Downing. New York, 1857.
Du Breuil, Cours Elémentaire d'Arboriculture. Paris, 1857.

Elliot, F. R., Fruit Book. New York, 1854.
——— Western Fruit Book. New York, 1859.
Ellis, Robert, Chemistry of Creation. London, 1850.
Encyclopædia Americana. Philadelphia, 1834.
Encyclopædia Britannica. 8th edition, Edinburgh, 1852–60.
English Cyclopædia. London, 1854.

Fentwanger, Lewis. Fermented Liquors. New York, 1858.
Fitch, Asa, Report on the Insects of the State of New York.
Flint, C., Agriculture of Massachusetts. Boston, 1858.
Floy, M., Guide to Orchard and Fruit Garden. New York, 1852.
Forsyth, Wm., Culture and Management of Fruit Trees. London, 1802.
French, Art of Distillation and Manufacture of Liquors. London, 1657.

Gardener's Chronicle. 19 vols., London, 1841, *continued.*
Gardener's Monthly. Philadelphia, 1859, *continued.*
Gardener's Monthly Volume. London, 1849.
Genesee Farmer. Rochester, N. Y. (*published monthly.*)
Gmelin, Leopold, Handbook of Chemistry. 12 vols., London, 1848–58.
Graham, Thomas, Elements of Chemistry. 2 vols., London, 1858.
Gray, Asa, Manual of Botany. New York, 1858.
——— Systematic and Structural Botany. New York, 1858.
Guide to Importers and Purchasers of Wines, with a Topographical Account of all the known Vineyards in the World. London, 1828.

Hannan, John, Economy of Waste Manures. London, 1844.
Harris, Joseph, Rural Annual, 1857, 8, 9. Rochester, N Y.
Harris, T. W., Report on the Insects of Massachusetts injurious to Vegetation. Cambridge, 1841.
Hoare, Clement, Treatise on the Grape Vine. New York, 1850.
Homans, J. S., Cyclopædia of Commerce. New York, 1858.
Hooker, Journal of Botany. 4 vols., London.
Horticulturist, 1846–1859.
Hovey, C. M., Magazine of Horticulture.

Jacques, Geo., Practical Treatise on Fruit Trees. Worcester, 1849.
Johnson, Geo. W., The Gardener. 3 vols., London, 1849.
——— Dictionary of Modern Gardening, edited bv Landreth. Philadelphia, 1857.
——— Principles of Practical Gardening. London, 1845.
Johnston, James F. W., Lectures on the Applications of Chemistry and Geology to Agriculture. New York, 1858.
——— Chemistry of Common Life. 2 vols., New York, 1855.
——— Notes on America. 2 vols., Boston, 18—.
Johnson, S. W., The Culture of the Vine. New Brunswick, N. J., 1806.
Jullien, Topographie de tous les Vignobles connus. Paris, 1816.

Kendrick, Wm., New American Orchardist. Boston, 1848.
Kollar, V., Treatise on Insects injurious to Farmers and Gardeners. London, 1840.

Ladrey, M. C., Chimie appliquée à la Viticulture et à l'Œnologie. Paris, 1857.
Liebig, J., Complete Works. Philadelphia, 1856.
——— Letters on Modern Agriculture. New York, 1859.
Liebig and Kopp, Annual Report on the Progress of Chemistry and the allied Sciences. London, 1847–1850.
Lindley, Theory and Practice of Horticulture. 2d edition, London, 1855.
——— the same, edited by A. J. Downing. New York, 1852.
Loudon J. C., Encyclopædia of Gardening. London, 1850.
——— Gardener's Magazine. 16 vols.

Manures, Practical Treatise on. Society for Diffusion of Useful Knowledge. London, 1830.
McCulloch, Remarks on the Art of Making Wine. London, 1817.
McIntosh, Charles, Book of the Garden. 2 vols., Edinburgh, 1855.
McMahon, Bernard, American Gardener's Calendar. Philadelphia, 1859.
——— the same, Philadelphia, 1806.
McMullen, Thomas, Hand-book of Wines. New York, 1853.
Meteorological Observations made in the State of New York from 1826 to 1850. Albany, 1855.
Meteorological Register, State of New York.
Miller, Philip, Gardener's Dictionary. Large folio, 1759.
——— the same, 4 vols., folio, 1807.
Morewood, Samuel, Essay on Meliorating Liquors. London, 1824.
Mulder, C. J., Chemistry of Wine. London, 1857.
——— Chemistry of Animal and Vegetable Physiology. Edinburgh, 1849.
Muspratt, Sheridan, Chemistry applied to Arts and Manufactures. Glasgow, 1858.

Natural History of the State of New York. 19 volumes, Albany.
Neil, Patrick, Practical Fruit, Flower and Kitchen Gardener's Companion. New York, 1856.
Northern Fruit Culturist.

Odart, Comte, Ampelographie Universelle. Paris, 1854.
——— Manuel de Vigneron.

Patent Office Reports. Washington, D. C., 1837–1858.
Pereira, Jonathan, Treatise on Food and Diet. London, 1844.
Perzoz, Nouveau Systême de Culture de la Vigne. Paris.
Philosophical Magazine (Tilloch's). 97 vols, London, 1798, *continued.*
Prince, W. R., Treatise on the Vine. New York, 1830.
——— Treatise on Horticulture. New York, 1828.
——— Pomological Manual. New York, 1832.

Quarterly Journal of Science and Art. 30 volumes, London, 1816-1830.

Redding, Cyrus, History and Description of Modern Wines. London, 1851.
Reemelin, C., Vine Dresser's Manual. New York, 1856.
Register of Rural Affairs. Albany, L. Tucker, 1855–1859.
Rendu, Ampelographie Française. Paris, 1857.
Rural New Yorker (*weekly*). Rochester, N. Y.

Schenck, P. A., Gardener's Text Book. Boston, 1852.
Schow, J. F., Earth, Plants and Man. London, 1852.
Silliman, B., American Journal of Science and Art. New Haven, Ct., 1819–1858.
Smeed, Wine Merchant's Manual. London, 1828.
Skinner, John S., Journal of Agriculture. 3 volumes, New York, 1848.
Speechly, William, Treatise on the Culture of the Vine, and the formation of Vineyards. 4to., London, 1790.
——— the same, 8vo., 1821.
Solly, Edward, Rural Chemistry. Philadelphia, 1852.
Somerville, Mary, Physical Geography. Philadelphia, 1853.
Stockhard, Julius C., Chemical Field Lectures. London, 1858.
Spooner, Alden, The Cultivation of American Grape Vines. Brooklyn, 1846.

Thayer, Albert, Principles of Agriculture. London, 1845.
Thomas, J. J., Fruit Culturist. Buffalo, N. Y., 1847.
——— the same, New York, 1857.
Transactions American Philosophical Society. Philadelphia, 1789.
Transactions of Society for Encouragement of Arts and Manufactures. London, 1783–1850.
Transactions of New York Agricultural Society. Albany, 1842–1858.
Transactions of New York Institute. Albany, 1841–1858.
Transactions of Royal Horticultural Society of London. 8 vols. 1824–1840.
Tucker, L., Register of Rural Affairs. Albany, 1855–1859.
Tull, Jethro. Horse Hoeing Husbandry. London 1829.

Ure, Andrew, Dictionary of Arts, Manufactures and Mines. New York, 1857.
——— Dictionary of Chemistry. Edinburgh, 1824.

Watson, American Home Garden. New York, 1859.
Webster, Encyclopædia of Domestic Economy. New York, 1856.

OPEN AIR GRAPE CULTURE.

CHAPTER I.

NATURAL AND CIVIL HISTORY OF THE VINE.

PROFANE history reaches not back to the time when man first planted a vineyard and made wine, and when we leave the sacred records, its first culture is shrouded in allegories, myths and fables.

The native country of the vine cannot be well ascertained. It occurs wild in Greece, Italy and even in the south of France. In Mingrelia, Georgia and the regions between Caucasus and Ararat and Taurus, it flourishes in extreme vigor and great abundance. And that it is indigenous to America, also, there can be no doubt, the apocryphal stories about its introduction by Sir W. Raleigh to the contrary notwithstanding.

Records of its culture are found in most of the poems and sculptures of antiquity. The shield of

Achilles represented a vine-gathering, and Herodotus and Theophrastus speak of the culture of the vine in Egypt; and on the very oldest Greek tombs are found pictures representing the vine harvest. Pliny enters fully into the natural history of the vine, and describes a variety with berries shaped like the finger,* while the second book of Virgil's Georgics forms no mean treatise on practical viticulture.

The generic name of the vine (*vitis*) is derived, according to some authors, from the Latin *vincire* to bind; according to others it comes from *viere*, to bend, alluding to the flexibility of its branches. Both these

* Most of the authors who have noticed this variety, suppose it to have been lost, but we have received from John Kolber, Esq., of New York, slips of a vine imported by him from Hungary, the fruit of which is described as being an inch and a half long and half an inch in diameter—a form which might easily be described by an imaginative writer as resembling a finger. In fact the native name is *Keckse, csocs* or *Goats teats*—an idea similar to that of the old philosopher, though not quite so elegant. We find also in several catalogues grapes called finger-grapes—synonyms of which are Cornichon Blanc, Cucumber Grape, Bec d'Oiseau (Bird's beak), Teta de Vacca (Cow's teats), Doigts Douzelle, etc., etc. Mr. Kolber has made earnest and praiseworthy efforts to introduce the hardier varieties of the vine from the hills of Hungary, and we are happy to learn that thus far, the results are exceedingly promising. It will take several years, however, to decide whether or not any foreign variety can be grown with success in this country, as most imported plants do well for a few years.

Latin words, however, are derived from a Greek word signifying to bind. Dr. Whittaker, in a work published in 1638, entitled, "The Tree of Human Life, or the Blood of the Grape," expresses his opinion that the name *vinum* is derived *a vi* from its strength, or, perhaps *quasi divinum*, because it is a species of the tree of life in Paradise.

The species of the genus vitis are numerous, though botanists are not agreed as to the distinctive differences, especially as between the European and American sorts. In France, Chaptal, when Minister of the Interior, caused 1,400 different varieties of the vine to be collected in the garden of the Luxembourg, and under his direction M. Champagny described as distinct 550 different kinds. Four American species have been usually numbered (some authors describe eight), though the varieties, more or less distinctly marked, probably exceed 300. To the number of the latter, however, there is no limit, as every seed *may* produce a new variety.

The vine lives to a great age and attains a great size. Pliny mentions a vine which had lived for 600 years, and in Italy, vineyards have continued in bearing for 300 years, while in some parts of that country, a vineyard of 100 years is still accounted young.

Its size, whether we regard the European or American varieties, is often very great. Speechly describes and figures a vine trained against a row of houses in Northallerton, Yorkshire, which covered a space of one hundred and thirty seven square yards, and had a stem three feet eleven inches in circumference at a short distance from the ground. No work on the grape vine would be complete without a mention of the great Hampton Court vine, from which George the Third once directed his gardener to cut one hundred dozen bunches of grapes, if so many were on the vine, and present them to the players of Drury-lane Theatre, who had greatly pleased him. The gardener not only cut off this number, but sent word to the king that he could cut off as many more without entirely stripping the vine. This vine was planted in 1769 and has a stem fourteen inches in girth, one branch extending nearly 200 feet.

In America, too, very large vines are to be found. The following is clipped from the "Alta Californian:"

"At Monticito, four miles from Santa Barbara, there is a grape vine, probably the largest in the world. Its dimensions and yield would be incredible, were it not that my informant is a man of veracity, and he spoke from personal observation. It is a single vine, the main stock being ten feet in circumference. It is trained upon a trellis sixty feet in diameter. My informant with another person counted 7000 bunches, and the estimated yield was 18,000 pounds of fruit. Can this be beaten? The only thing that surprised me in the relation of my friend was that any person in Santa Barbara should have displayed the energy necessary to build the trellis for this noble vine."

In the "Horticulturist" for October, 1858, a vine growing near Burlington, New Jersey, is described as follows: "In May last it was measured with the following result: Two feet from the ground it measured 6 feet, 2½ inches in girth; four feet high it is about 6 inches less; it there divides into two branches, the largest of which is 3 feet, 3 inches in girth, and the smallest is 3 inches. The largest of the trees which the vine covers is 10 feet in circumference at two feet from the ground. The vine is very much decayed, but still puts forth leaves and young shoots. It has never borne a grape in the memory of a lady now 98 years old and who has lived her long life within sight, or nearly so, of this gigantic production, and to whom it was a wonder in her youth. The largest tree is a black oak, the others are black, or sour gum. On pacing the circumference covered by the branches, it was found to exceed 100 feet.

"This vine grows near a springy soil, or upland, its roots, no doubt, penetrating to the water. May not this teach us a lesson, to give the rootlets, wherever it is possible, access to a spring, or running water? It may be a question, too, whether we do not cut our vines too much. I have observed frequently in England that a whole house was devoted to a single vine, generally of the Black Hamburgh, and I think they uniformly bore the finest grapes. To carry a single vine over a large grapery would, of course, require years of judicious trimming and management."

The bunches and berries also have been known to attain a very great size. In the south of France instances are known of bunches attaining a weight of eight or ten pounds; travellers in Syria mention bunches weighing 17 lbs.; and we all remember the enormous clusters which the Jewish spies brought back from the promised land. Even at the present day the grapes of Damascus frequently weigh 25 pounds to the bunch.

With all the vigor and fruitfulness evinced by such instances it is no wonder that the culture of the vine should prove profitable and certain. At the meeting of the Fruit Growers' Society for western New York, held in the city of Rochester in 1859, S. H. Ainsworth made some statements as to the actual products of several vineyards, showing that from $1000 to $1500 had been realized from an acre of Isabella grapes. Mr. Rush, of East Bloomfield, had 100 vines on one-third of an acre, from which he picked 4000 lbs., which he sold for $500, or at the rate of 12½ cts. per pound. None reported a less profit than $500 per acre.

From the very first settlement of America the vine attracted the attention of the colonists, and efforts were made both to introduce the finer European varieties and to cultivate the native sorts. Even as early as 1564, wine was made from the native grape in Florida, though, of course, in small quantity.

The earliest attempt to establish a vineyard in the British North American colonies was by the "London Company" in Virginia prior to 1620. By the year 1630, the prospects were sufficiently favorable to warrant the importation of several French *vignerons*, who, it was alleged, ruined them by bad management. Wine was also made in Virginia in 1647, and in 1651 premiums were offered for its production. On the authority of Beverley, who wrote prior to 1722, there were vineyards in that colony which produced 750 gallons a year.

In 1664, Col. Richard Nicolls, the first English governor of New York, granted to Paul Richards of the city of New York the privilege of making and selling wine free of all duty or impost, Richards having been the first to enter upon the culture of the vine on a large scale. It was also enacted that every person who should during the succeeding thirty years set out a vineyard should pay to Richards five shillings for every acre of vines so set out. We have been unable, however, to find any account of his success or failure, and the probability is, that after a short time the enterprise was abandoned. A gentleman in Hoboken, also, had a fine vineyard which after a little time fell into decay.

Beauchamp Plantagenet, in his "Description of the Province of New Albion," published in London in 1648, states that the English settlers in Uvedale (now Delaware) had vines running on mulberry and sassafras trees, and that there were four kinds of grapes. "The first is the Tholouse Muscat, sweet scented; the second, the great fox and thick grape, after five moneths reaped, being boyled, and salted, and well-fined is a strong red Xeres; the third, a light claret; the fourth, a white grape, creeps on the land maketh a pure, gold-colored wine. Tennis Pale, the Frenchmen, of these four made eight sorts of excellent wine; and of the Muscat, acute boyled, that the second draught will fox (intoxicate) a reasonable pate four moneths old; and here may be gathered and made two hundred tun in the vintage moneth, and replanted will mend."

In 1683, William Penn attempted to establish a vineyard near Philadelphia, but without success. The same result attended the efforts of Andrew Doré in 1685, but after some years, Mr. Tasker, of Maryland, and Mr. Antil, of Shrewsbury, N.J., seem to have succeeded to a certain extent. Mr. Antil wrote an excellent article on the culture of the grape and the manufacture of wine, which may be found in the first volume of the "Transactions of the American Philosophical Society," published in 1771. In this article, Mr. Antil describes only foreign varieties, from which it is to be inferred that he cultivated them chiefly, if not solely.

In 1769, the French settlers in Illinois made one hundred and ten hogsheads of strong wine from native grapes.

In 1793, Peter Legaux, a French gentleman, obtained of the legislature of Pennsylvania the incorporation of a company for cultivating the vine. They purchased a farm at Spring-mill, Montgomery county, thirteen miles from Philadelphia, on the Schuylkill. For one year only were prospects favorable; divisions and dissensions arose; the stockholders sold out in disgust, and the vineyard went to ruin.

At Harmony, near Pittsburg, a vineyard of ten acres was planted and cultivated by Frederick Rapp and his associates from Germany. They afterward removed to another Harmony in Indiana, on the east bank of the Wabash, where they continued the cultivation of wine and silk for many years.

A Swiss colony settled about 1790 in Jessamin county, Kentucky, and raised a fund of ten thousand dollars for the express purpose of forming a vineyard. Their first attempts failed, they having cultivated the foreign vine. In 1801, they removed to a spot which they called Vevay, in Switzerland County, Indiana, on the Ohio River, forty-five miles below Cincinnati. Here they planted native vines and met with some success. But, after forty years' experience, they consider our climate and soil inferior to those of Switzerland, as they claim that they can there make a gallon of wine from ten pounds of grapes while here twelve pounds are required. Their vineyards have now, we believe, nearly disappeared.

But the great turning point of vine culture in America was when the Catawba grape was introduced by Major Adlum, of Georgetown, D. C., who considered that in so doing he conferred a greater benefit upon the American nation than he would have done by paying off the national debt.

We could have wished to give an accurate view of the present state of the vine culture of this country, but the best works which we have been able to consult are very imperfect in this respect, and we believe that we have examined all the more important ones. Want of time has prevented us from instituting a special correspondence on this subject. We can therefore only say that it never at any period presented a more flourishing aspect than it does at the present day.

Of the future prospects of grape culture, of its extent, and of its influences, it would be difficult to speak. But we feel assured that, whether in the form of wine or of fruit, the produce of the vine cannot fail to do much good in this country—not the least of its benefits being that it will afford those with small capital an easy and pleasant mode of securing a competency.

Another point in this aspect of grape culture, and one in which we have strong confidence and ardent hope, is the employment which it promises to afford to women. We are none of those who would desire to see woman rendered independent of man, for we well know to what a miserable condition man would come if rendered independent of woman, and it is a poor rule that will not work both ways.

But we cannot shut our eyes to the fact that there are vast multitudes of women whose labor receives no adequate remuneration—who make shirts at the rate of five cents apiece, and then often get cheated out of their pay. Now, if some of our large-hearted, as well as large-worded, philanthropists would procure a few acres of land in some proper locality, and after having it well trenched or subsoil ploughed, would let it out in half, or even quarter acre lots to industrious women with a view to their establishing vineyard plots, we think that after the first two years such an allotment of half an acre ought to yield its tenant from $250 to $400 per year, from which, after paying a good round rent, they might retain more than they can now make at any other employment within their reach. And let it not be said that the culture of the soil is unsuited to the sphere of woman. We rather think that Eve was more of a gardener than shirt-maker before she "brought sin into the world and all our woe;" and those who think gardening unsuited to woman are referred to Loudon's remarks in the "Gardener's Magazine," where he recommends it to his fair countrywomen instead of the ball-room and the dance.

We shall not stultify ourselves with referring to Indian and European savages, who make the women do all the hard work, even though women are there found equal to the roughest agricultural labor. But in vine culture, after the first great effort has been made to get the soil suitably prepared, there is really little hard work to be done. Even hoeing does not require more strength than washing and scrubbing; and pruning, trimming, and gathering the fruit are not above the strength of our weakest females; and we promise them that if they undertake it they will soon acquire the necessary health and strength. All that we can say is that we hope ere long to see the experiment tried, and nothing would afford us greater pleasure than to give a lecture on vine culture, with experimental illustrations, to such a society of women, and tell them all we know about raising good grapes; and we think we can point to others who are not only competent but willing to assist in the good work —thus rendering the objection that "women don't know how" of no avail. But even if no such experiment should be tried, we feel confident that the thousands of acres which will be devoted to vine culture during the next few years will not be cultivated without affording abundant work for women.

CHAPTER II.

SOIL, SITUATION AND ASPECT.

SOIL.—The vine will grow in almost any situation, and reach a large size and exhibit luxurious vegetation under conditions apparently the most unfavorable; but if healthy vines and fine fruit be desired, it is necessary to choose a soil where the roots can ramble freely, find plenty of nutriment and be safe from stagnant water and its accompanying cold, sour subsoil. One of the largest vines in the country grows in a swamp in New Jersey, and a vine has been known to grow vigorously from a cleft in an old wall twenty feet from the ground. But these are by no means examples to be imitated in practice where we have the power of selecting the site of our garden or vineyard, though they afford encouragement to the amateur who is compelled to make use of an inferior location.

The opinion of good grape culturists is that any soil which will grow *good* Indian corn is suitable for grapes. Others describe a soil adapted to the culture of the vine as one which will grow good winter wheat without the plants being thrown out of the ground in winter.

Downing recommends a "strong loamy or gravelly soil—limestone soils being usually the best." And in another place he gives it as his opinion that "all that can be said of a soil for grape culture is that it be light, rich and dry." G. W. Johnson thinks a light, sandy loam the best. And Buchannan, who may be safely taken as the representative of the Cincinnati vine growers, recommends a dry, calcareous loam with a porous subsoil. At the recent meeting of the Fruit Growers' Society of western New York, Dr. Farley stated that his best grapes had been raised on a clay soil, and that in this matter his opinion in regard to the soil best adapted to the culture of grapes had undergone some change.

It will thus be perceived that the opinions of our best horticulturists vary a little, but we believe that this variation is mere adaptation to the different modes of growth and training adopted by the various cultivators. The purpose for which the grapes are raised —that is whether for wine or for the table—ought also to have a material influence in directing our choice of a soil.

When the object is to manufacture wine, the vines require to be kept within moderate bounds; all rankness of vegetation must be carefully avoided, and consequently the soil must be light, rich, porous and dry, and if calcareous so much the better.

On the other hand, where high saccharine qualities are not so much desired as abundance of grapes of agreeable flavor, the vines will succeed better and produce more certain crops if allowed a greater extent of growth, and in this case they will bear a heavier and richer soil—in some cases (as in growing Isabella and Diana grapes for the table) even preferring a clay soil well drained and cultivated and highly manured.

That this view is correct may be easily proved by referring to well-known examples both in Europe and in this country. Thus in the Arriege in France a rich wine like Tokay, is obtained from mountain sides covered with large stones as if the cultivators had left all to nature. In Italy and Sicily the best wines are grown amongst the rubbish of volcanoes. "Good rich soils," says Redding, "never produce even tolerable wines."

On the other hand the rich Chasselas de Fontainebleau table grapes are produced by vines planted in cold and heavy soil, well manured. And he who desires to find *rich soil* should examine the vine borders of the English hot-house grape-growers. Allen, one of our most successful grape-growers recommends a border of the richest kind. So does Chorlton, and such we believe to be the practice of all our successful cultivators of the grape under glass. The celebrated vine at Hampton Court revels in the luxury of an old sewer, and instances have come under our own observation where the proximity of a vine to a cesspool caused the production of large quantities of most excellent grapes. In France, the application of night-soil and sewerage to the vineyards has in all cases injured the quality of the wine. That such would have been the case, however, if the French vignerons had acted upon correct principles in the application of these powerful stimulants, we are scarcely prepared to believe. And we have no doubt but that by judicious management and a careful observance of the laws of nature one of the greatest achievements in vine culture may yet be effected, viz., the union of vigorous vegetation and stimulating manures with the production of good wine. But so far as present experience extends the soil for a vineyard must be light and not *too highly manured*—and in all cases whether the object of culture be wine or table grapes the subsoil must be warm and loose. Cold borders are very prejudicial to the roots of the vine, and are supposed to be an efficient cause of the *shanking* of the grapes. It would appear from an inspection of the subjoined tables that this desired warmth might be secured to the surface soil at least by plentiful addition of lime and any black mold or charcoal.

Maximum Temperatures of the various Earths Exposed to the Sun. By Schubler.

KINDS OF EARTH.	Maximum Temperature of the superior layer, the mean temperature of the ambient air being 77 degrees F.	
	Moist Earth.	Dry Earth.
	Degrees.	Degrees.
Silicious sand, yellowish grey, ...	99.05	112.55
Calcareous sand, whitish grey,....	99.10	112.10
Argillaceous earth, yellowish grey,	99.28	112.32
Calcareous earth, white, ...	96.13	109.40
Mold, blackish grey,............	103.55	117.27
Garden earth, blackish grey,.....	99.50	113.45

Table of Retention of Heat. By Becquerel.

KIND OF EARTH.	Capacity for heat, that of Calcareous sand being 100.	Time required by 18 feet cube of earth to cool from 144.5 to 70.2, the temperature of the surrounding air being 61°.2.
		Hours.
Calcareous sand, ...	100	3.30
Silicious sand,........	95.6	3.27
Argillaceous earth,....	68.4	2.24
Calcareous earth,.....	61.8	2.10
Mold, ...	49	1.43

From these tables it will be seen that black mold receives or absorbs heat most rapidly, but parts with it in the shortest space of time also, and that for receiving and retaining heat, dark colored, calcareous earth is by far the most efficient. Good silicious sand comes next in order, and hence we conceive that a soil composed chiefly of calcareous and silicious sand, with a sufficient amount of charcoal or mold to give it a dark color, would prove one of the best for grapes.

Such are the general points deserving of consideration. Those desirous of studying more minutely the influence of the chemical constitution of the soil upon vines growing therein will find an interesting and valuable résumé of the subject in M. Ladrey's "Chimie appliqué à la Viticulture," whose general remarks on this point are so much in unison with our own experience and observation that we are tempted to translate them.

"If now we examine the series of different soils devoted to the culture of the vine in France and in other countries, we shall find this plant cultivated in soils the most diverse, not only as regards their natures (nature evidently alluding to physical constitution—*Trans.*)—but also their chemical composition. All soils appear suited to the culture of the vine, and there are none, unless those absolutely barren, in which this plant may not grow and develop itself. Thus the vine requires but little fertility in the soil, it covers a great space of land which would be unsuited to any other culture, and in order to give an idea of this, we may cite the ancient regulations of Provence which prohibited the planting of the vine until inquiry had been made as to the sterility of the soil, and the permission of the intendant of the province had been obtained.

But if the vine can grow in all soils it behaves very differently in each of them. In strong, argillaceous, rich soils, it will acquire a great vigor of vegetation, the wood is largely developed, the product is abundant; on the contrary, in soils poor, light and dry, the vine is less robust, more delicate; it requires a culture well contrived as to even the most minute details, and the product is much less in quantity.

"In general, if in any locality the vegetation of the vine be more rich as the soil is more fertile, we observe by the side of this result that the nature and quality of the product is consequently in an inverse ratio. In heavy land the vine is well developed and furnishes abundant return; in a light soil it gives less and the product is of higher quality."

Situation.—The situation of a vineyard should be elevated, but not too high, otherwise the vines will not only be exposed to high winds and their concomitant evils, but will also be subjected to a lower temperature. On this latter point, but little is known—at least not enough to enable us in all cases to reconcile the anomalies which occur. Enough is known, however, to cause us to avoid the tops of hills and the bottoms of valleys, and it may be worth our while to consider a few of the principles which regulate temperature in these situations. During the night, the cold air, being heavy, settles down into the valleys and hollows, thus producing in such locations a temperature several degrees lower than is found on the sides of the adjacent hills. And no influence is then at work to disturb this state of things, for the earth itself is becoming rapidly cooled by radiation; and if a small quantity of the air should become warmed by contact with it, it immediately ascends, and cool air takes its place.

At daybreak, however, an agency is introduced which reverses this condition of things. Then the dense air in the valleys concentrates and absorbs the heat of the sun's rays and increases their effect upon the soil, which in turn imparts heat to the stratum of air lying next it. This lower stratum of air being warmed and consequently rendered much lighter than the colder portion above it, it ascends, but as it rises it also expands still more, which in some measure compensates for the heat which it received from the earth. The same process keeps going on until night comes, when the lower stratum of air being no longer warmed it no longer ascends, and the colder and heavier air again accumulates in the valleys. Thus it will be seen, that during the night the air in the valleys is colder than that in other places, while the reverse is the case during the day. The stillness of the air in valleys and sheltered situations also contributes to this result in a remarkable degree.

Now it is obvious, that if for any fruit tree, the air in the valleys should be sufficiently cold to kill the buds, no orchard could succeed. And if, on the other hand, sufficient light and heat to ripen the fuit could not be found on the hill-tops, such situations also would be unavailable.

Nor is the mere existence of such extremes of temperature the worst evil. The destructive influence of a hot sun upon frozen vegetation is well known, and in low valleys, the circumstances are such as to give the greatest effect to this adverse influence. For not only are the plants chilled by the extra cold night-air, they are also completely protected from the rays of the sun, until it has attained a greater power than it usually exerts at its first appearance upon plants in more exposed situations. And then, owing to the dense atmosphere through which they pass, the rays strike suddenly with concentrated energy so as to thaw the buds with a rapidity completely destructive to their vitality. In such situations also, the soil is usually very deep and rich, producing a vigorous though succulent growth which is unable to withstand the influences above detailed. All experience bears out the practical value of these principles. Thus, in Italy, where the country is undulating and very much broken, all good wines are grown on the hill-sides. Hence Virgil tells us

> . . . . . "denique apertos
> Bacchus amat colles,"*

and modern experience bears out the ancient saw, though it does not follow, however, that plains will not produce good wine-making grapes, provided they be of sufficient extent to obviate the evils just described. The fine wines of the Gironde in France, and Châtaux Margaux, Lafitte and Latour, are grown on the plains.

Aspect—Exposure.—The aspect which is best adapted to the growth of grapes will, of course, depend upon influences, some of which at least, are liable to vary, as the keenest and most destructive winds may come from different quarters in different places—a very slight geographical change sometimes making

* The force of this saying is lost by adopting Mr. Redding's translation "Bacchus loves the hills." Davidson gives the whole, "Bacchus loves the open hills"—which is better. But the true meaning "Bacchus loves the open little hills" coincides perfectly with experience and with the principles above set forth.

an important difference in this respect, owing to peculiar topographical features. Thus a range of hills or a belt of woods, may so deflect the prevailing winds, as to completely change the condition of two localities situated within even a very short distance of each other.

In general, it will be found necessary to secure protection on the west, north and northeast. This may be afforded either by natural local features, as by a range of hills, or it may be derived from artificial sources, as woods or fences. No defence is better than a good belt of Norway spruce, and if they form a crescent in which the vineyard is embowered, but little danger need be apprehended from violent winds. Even high fences, which may be single, double or triple, afford ample protection in ordinary cases, and as trees, even of the fastest growing kind, take a considerable time before they give sufficient protection, many will prefer the fence. We are therefore tempted to extract from the "Horticulturist" for August, 1847, Downing's description of the method by which Frederic Tudor, Esq., has converted the naked promontory of Nahant into a luxuriant garden.

"To appreciate the difficulties with which this gentleman had to contend, or as we might more properly say, which stimulated all his efforts, we must recall to mind that, frequently, in high winds, the salt spray drives over the whole of Nahant; that until Mr. Tudor began his improvements, not even a bush grew naturally on the whole of its area; and that the east winds which blew from the Atlantic in the spring are sufficient to render all gardening possibilities in the usual way nearly as chimerical as cul tivating the volcanoes of the moon. Mr. Tudor's residence there, now, is a curious and striking illustration of the triumph of art over nature.

"Of course, even the idea of a place worthy of the name of a garden in this bald, sea girt cape, was out of the question, unless some mode of overcoming the violence of the gales and the bad effect of the salt spray could be devised. The plan Mr. Tudor has adopted is, we believe, original with him, and is at once extremely simple and perfectly effective.

* * * * * * * *

"It consists merely of two, or at most three parallel rows of high open fences, made of rough slats or palings, nailed in the common vertical manner, about three inches wide, and a space of a couple of inches left between them. These paling fences are about 16 feet high, and usually form a double row (on the most exposed side, a triple row) round the whole garden. The distance between that on the outer boundary and the next interior one is about four feet. The garden is also intersected here and there by tall trellis fences of the same kind, all of which help to increase the shelter, while some of those in the interior serve as frames for training trees upon.

"The effect of this double or triple barrier of high paling is marvellous; although like a common paling, apparently open and permitting the wind a free passage, yet in practice it is found entirely to rob the gales of their violence and their saltness. To use Mr. Tudor's words, 'it completely sifts the air.' After great storms, when the outer barrier will be found covered with a coating of salt, the foliage in the garden is entirely uninjured. It acts, in short, like a rustic veil, that admits just so much of the air, and in such a manner as most to promote the growth of the trees, while it breaks and wards off all the deleterious influences of a genuine ocean breeze, so pernicious to tender leaves and shoots.'

* * * * * * * *

"It is worthy of record, among the results of Mr. Tudor's culture, that two years after the principal plantation of his fruit trees was made, he carried off the second prize for pears at the annual exhibition of the Massachusetts Horticultural Society, among dozens of zealous competitors, and with the fruit most carefully grown in that vicinity."

Of the necessity for shelter under circumstances far less desperate than those at Nahant, no good horticulturist has any doubt. Even the oak-tree has been proved by a well directed series of experiments, to be benefited by shelter in the comparatively mild climate of England. For the rationale of the evil effects of wind on plants in general, we must refer the reader to Lindley's "Theory and Practice of Horticulture." The following cases are detailed by Hoare:

"Many instances might be circumstantially detailed of the injurious effects of wind upon established vines during their summer's growth; two, however, of recent occurrence will perhaps suffice.

"On the eleventh of June, 1833, a strong wind sprang up early in the morning from the west, and increased in force till noon, when it blew quite a gale and continued to do so throughout the day. It slackened a little during the night, and gradually decreased in violence the next day, dying entirely away in the evening.

"The effects of this wind on a vine of the White Muscadine sort, trained on a wall having a western aspect, were carefully observed. It had on a full crop of fruit and a good supply of fine young bearing shoots, and was altogether in a most thriving condition. Such, however, were the injurious effects of the wind in dissipating all the accumulated secretions of the foliage, and then closing, almost hermetically, its pores, and thereby totally deranging the vital functions of the plant, that although in the height of the growing season, not the slightest appearance of renewed vegetation could be discerned in any part of its leaves, shoots or fruit, until the third day of July, or twenty-two days afterward. It never produced another inch of good bearing wood throughout the remainder of the season, but lingered in a very weak and sickly condition; and the fruit which had been previously estimated at ninety pounds' weight, did not exceed fifty-five pounds when gathered, and that of a very inferior description in point of flavor and size of berry. Its leaves, also, having been thus crippled, were shed prematurely a month before their natural time, and hence the deficiency in the flavor and size of the grapes.

"The other instance, which happened shortly afterward, is still more decisive. On the 30th of August following, about eight o'clock in the evening, a strong wind began to blow from the southwest, accompanied with heavy rain. At nine it blew violently, and continued to do so until noon the next day. It then slackened, and then veering to the northwest, died away some time during the following night.

"The full force of this wind fell on a remarkably fine black Hamburg vine, trained on a wall having a southwestern aspect, and its effects were therefore proportionately destructive. Many of the principal branches were torn so completely from their fastenings that their extremities swept the ground. The bunches of fruit were knocked about, and portions of them, as well as single berries, lay scattered on the ground in every direction. On the fruit, however, that survived the wreck, the effects of the wind were remarkable. It must be stated that the wall on which the vine is trained, is ten feet high, and is so situated that to the height of about three feet from the ground the wind had but little power over it, its force being broken by an outer wall standing at a little distance off in front of it. On the lower part of the wall so protected, the grapes not having been much injured, began to change their color and ripen about the twentieth of September, and on the twelfth of October every berry was perfectly matured, while all those that remained on the vine above three feet from the ground, were, on the first of November, as green and hard as on the thirtieth of August, when the high wind occurred. Shortly afterward these began to change their color, and ultimately ripened tolerably well by the first week in December. Thus, solely through the effects of a strong wind, there were to be seen at the same time, on the same branches of this vine, and within nine inches of each other, bunches of grapes, the lowermost of which were perfectly ripe, while the uppermost were quite green and hard, and not within seven weeks of reaching the same state of maturity.

"These facts, which might be multiplied indefinitely, sufficiently show the injurious effects of strong winds, and the necessity of protecting vines as much as possible from their destructive consequences."

But although there can be no doubt as to the evil effects of wind storms, it must be borne in mind that ventilation, and even motion, are essential to the health and growth of the vine. Experiments made by Andrew Knight, show that young trees tied to stakes so as to prevent all motion, do not increase in size as much as those left to the free action of wind. Hence, perhaps, one reason why wire is to be preferred to wood for the cross slats of trellises. In the northern States, however, we in general have wind enough for all useful purposes. But in view of these facts, we would rest content with shelter outside of the vineyard, and unless in very exposed situations we would not deem it advisable to place either trees or fences amongst the vines.

But while we can guard against wind and storms by belts of woods or high fences, there are other influences which we cannot thus alter. Chiefly among these is the exposure of the sun's rays.

Exposure is, in general, derived from one or both of two causes. First, the inclination of the ground, and, secondly, its openness and freedom from overshadowing influences. A wall is a good illustration of the latter—the north side having a northern exposure, and causing fruit planted against it to ripen at a much later period than that planted on the south side, which has a southern exposure. The little raised mounds or flower-beds, to be found in every garden, exhibit the influence exerted by the inclination of the earth—the vegetation on the south side being usually some days earlier than that on the north.

For vineyards, the best exposure is undoubtedly a southern one, slightly inclined toward the east, or at least fully protected from the west, and also from the early morning rays. "It has often been observed that woods or thick trees, buildings, high, broad fences, or steep hills, on the east side of peach orchards, protect the crop. Hence the erroneous opinion, that it is the east winds which do the damage. It is the sunshine upon the frozen buds which destroys them; hence a clouded sky, after a clear frosty night, by preventing sudden thawing, sometimes saves a crop. Covering trees of rare kinds with mats, to shade them from the morning sun, after an intensely frosty night, might sometimes be highly beneficial." (Thomas.)

In this connection, it may be proper to consider

the best direction for the trellises on which the vines are trained. We have often seen a north and south direction advised under the idea that the vines thus receive the sun's rays for a longer time. But the evils attached to this plan are great and insurmountable. In the first place, the vines receive the full force of the early morning sun which, striking the young leaves while still cold, and it may be partially frozen, is productive of the most injurious effects. Then as the day progresses toward noon, the vines are so shaded as not to receive the amount of heat, which they would gladly enjoy at that time, while toward evening again their excitability is greatly increased and is kept up until the last moment, instead of the exciting influence being quietly withdrawn as it ought to be.

But if we give our trellis a direction from east to west, instead of from north to south, the vines will expose but a small surface to the first rays of the sun which will thus warm them gradually, until it attains its meridian splendor, when it will exert its full power and then gradually decline until evening, when everything will gradually cool down. Sudden changes are thus avoided, and the full power of the sun is secured in the ripening of the grapes.

Intimately connected with the foregoing subjects, are the laws which regulate the influence of temperature upon vegetation. These are stated by M. De Candolle, as follows:

1. All other things being equal, the power of each plant and of each part of a plant, to resist extremes of temperature is in the inverse ratio of the quantity of water they contain.

2. The power of plants to resist extremes of temperature is directly in proportion to the viscidity of their fluids.

3. The power of plants to resist cold is in the inverse ratio of the rapidity with which their fluids circulate.

4. The liability to freeze, of the fluids contained in plants, is greater in proportion to the size of the cells.

5. The power of plants to resist extremes of temperature is in a direct proportion to the quantity of confined air which the structure of their organs give them the means of retaining in the more delicate parts.

6. The power of plants to resist extremes of temperature is in direct proportion to the capability which the roots possess of absorbing sap less exposed to the external influence of the atmosphere and the sun.

From this it will be obvious that all rank growth and succulent vegetation should be avoided where the desired object is to obtain hardy vines.

CHAPTER III.

PREPARATION OF THE SOIL AND FORMATION OF VINE BORDERS.

Having selected a proper site for a vineyard, the next step will be to prepare the soil for the reception of the young vines. It is rarely if ever that ground can be found in a condition fit to plant a vineyard without thorough and extensive improvements, and unless it be in proper order our hopes of success will end in failure and disappointment.

In our remarks on soil it was stated that one absolute necessity is a *dry* subsoil. No other good qualities can compensate for the want of this, and in most cases it is only to be obtained by thorough *draining.*

The first great evil obviated by thorough draining is the existence of stagnant water beneath the surface. It is a saying amongst vine-dressers that "the vine cannot bear wet feet." And nothing can be more true. If the roots be exposed to stagnant water they will become diseased and die off, thus giving rise to weak and ill-ripened though sometimes succulent growth, and hence causing the vine to suffer from the attacks of disease and insects. The grapes, too, will not ripen well, but will remain sour and ill-flavored.

M. Gasparin gives the following observations with regard to the influence which a dry or a moist soil exerts upon the grape: "Other things being equal, we obtain grapes which contain much sugar and little acid from vines grown in a dry soil; more free acid in a moist soil, and much acid, albumen and mucilage with little sugar in a soil which is absolutely wet."

Another advantage consists in the fact that well-drained land always possesses a higher temperature than that which is wet. This difference amounts to 10° to 12° Fah. and is accounted for by the rapid absorption of heat by the water as it becomes converted into vapor. During this process, too, it is probable that the nascent vapor robs the earth of a portion of the ammonia and gases which it would have separated from the water and retained if it had acted as a filter and the water had passed off by the drains. But however this may be, its effect on temperature is such that Johnson regards thorough draining as equal to a change of climate.

But not only does draining enable the soil to filter all the water which descends upon it, retaining its ammonia, gases and even salts; it is probable that by these means the excrementitious matters discharged by plants, as well as other noxious bodies are washed out of the subsoil or decomposed by contact with the air which penetrates along with the water. In the case of oxide of iron it is probable that a very beneficial effect results from its conversion from the protoxide to the peroxide by means of this influence.

But a change in the chemical constitution and action of the soil is not the only effect of this operation; a no less marked alteration is produced in its mechanical character—heavy lands being rendered light, porous and permeable to the roots of tender plants.

It is unnecessary here to give minute directions for performing such a well-known operation, so we shall merely refer our readers to some of the numerous treatises on that subject. An excellent article on the theory and practice of draining will be found in the "Rural Annual" for 1859 published at the office of the "Genesee Farmer," Rochester, N. Y.

We may state, however, that in laying drains for a vineyard, it should be borne in mind that after the vines are planted it will be almost impossible to get at the drains in case of accident, without serious detriment to the plants. It will, therefore, be well to construct them in the most substantial manner and also to arrange them so that they will not lie immediately under any of the rows of vines. If they are *between* the rows it will not be so difficult to get at them as if they lay directly beneath the plants.

The next great requisite in a soil for the culture of the vine is depth. Ordinary soils of from eight to ten inches are by no means deep enough. Twenty inches is the least depth to be relied upon, and, if very favorable results are desired, it should be made three feet. The subsoil to this depth should be thoroughly loosened, and, unless its quality is very inferior, it may be well to mix it with the surface soil—adding at the same time a good supply of manure or compost. We are aware that some horticulturists object to bringing up the subsoil, but we incline to the belief that if it is of such a character as to produce much injury, the site is unfit for a vineyard. When the subsoil is light (except it be pure sand) no harm can result. If it be pure sand, however, it had better remain where it is unless a sufficiency of clay can be found to mix with it. If, on the other hand, it be so clayey as to hermetically seal up the vine borders, we should prefer to let it remain under. But, if possible, a site should be selected where a good depth of tolerable soil may be obtained either naturally or by proper effort.

The advantages incident to depth in ordinary cases consist in the roots being placed alike beyond the extreme heat of summer and the severe cold of winter. Consequently they do not suffer from drought, and are able at once to enter upon their duties in the spring.

For table grapes, we doubt whether the soil can be too deep or rich—not meaning by the latter term, however, saturated with *undecomposed* organic matter. But observation leads us to doubt the propriety of carrying these features to an extreme in the case of closely-trimmed vines cultivated for wine. It is true that the Western authors (Remelin, Buchannan, etc.—some of them Europeans) advocate this depth and richness. But, if our memory does not deceive us, some of Mr. Longworth's tenants who have not pursued the most thorough system of cultivation have occasionally escaped evils to which their more *skillful* and hard-working brethren have been exposed. And perhaps a solution of this mystery may be found above, notwithstanding Mr. Longworth naïvely tells us that he cannot believe that nature ever favors the indolent. Our own experience in this particular department is not sufficient to warrant us in pronouncing a decided opinion on the subject; but the principles of physiology would lead us to believe that if the roots of vines are planted in a deep and rich soil the branches must be allowed corresponding elbow room. If we desire to keep a vigorous plant down we must starve and curtail its roots as well as use the pruning-knife on its branches.

There are two methods of deepening a soil, viz: by the subsoil plough and by trenching with the spade. Both these operations are too well known to require a minute description, though in regard to the latter there are so many and such contradictory directions given in books that we may be pardoned a few remarks in relation thereto.

In order properly to trench a piece of ground the directions given by Loudon are as explicit and judicious as possible. "Trenching is a mode of pulverizing and mixing the soil, or of pulverizing and changing its surface to a greater depth than can be done by the spade alone. For trenching with a view to pulverizing and changing the surface, a trench is formed like the furrow in digging, but two or more times wider and deeper; the plot or piece to be trenched is next marked off with the line into parallel strips of this width; and beginning at one of these, the operator digs or picks the surface stratum, and throws it in the bottom of the trench. Having completed with the shovel the removal of the surface stratum, a second, third or fourth, according to the depth of the soil and other circumstances, is removed in the same way; and thus, when the operation is completed, the position of the different strata is exactly the reverse of what they were before. In trenching with a view to mixture and pulverization, all that is necessary is to open, at one corner of the plot, a trench or excavation of the desired depth, 3 or 4 feet broad, and 6 or 8 feet long. Then proceed to fill the excavation from one end by working out a similar one. In this way proceed across the piece to be trenched, and then return, and so on in parallel courses to the end of the plot, observing that the face or position of the moved soil in the trench must always be that of a slope, in order that whatever is thrown there may be mixed and not deposited in regular layers as in the other case. To effect this most completely, the operator should always stand in the bottom of the trench, and first picking down and mixing the materials, from the solid side, should next take them up with the shovel, and throw them on the slope or face of the moved soil, keeping a distinct space of two or three feet between them. For want of attention to this, in trenching new soils for gardens and plantations, it may be truly said that half the benefit derivable from the operation is lost."

A more expeditious method of mixing the soil, and one which varies but slightly from the ordinary system, consists in cutting down the bank in successive sections so as to produce theoretically a series of layers of soil and subsoil, but in reality a most intimate mixture of the two. This is best accomplished by opening a very wide trench—say from four to six feet wide. Then throw the top spit off a bank of the same width into the bottom of the trench so as to insure the burial of all insects, seeds, and weeds; cut a width of from six to fifteen inches of the remaining portion of the bank completely down to the bottom, and spread the soil so obtained in a thin layer over the spit formerly thrown in. Then cut down another six to fifteen inches in the same manner, proceeding thus until the whole bank has been cut down and used to fill up the trench. It will now be found that, with the exception of the extreme top spit which is placed at the bottom for very good reasons, the whole soil is sufficiently mixed for all practical purposes.

Another mode of trenching—called bastard trenching—is thus described by a writer in the "Gardener's Chronicle:" "Open a trench two feet and a half, or a yard wide, one full spit and the shovelling deep, and wheel the soil from it to where it is intended to finish the piece; then put in the dung and dig it in with the bottom spit in the trench; then fill up this trench with the top spit, etc., of the second, treating it in like manner, and so on. The advantages of this plan of working the soil are, the good soil is retained at the top—an important consideration where the soil is poor or bad; the bottom soil is enriched and loosened for the penetration and nourishment of the roots, and allowing them to descend deeper, they are not so liable to suffer from drought in summer; strong soil is rendered capable of absorbing more moisture, and yet remains drier at the surface by the water passing down more rapidly to the subsoil, and it insures a more thorough shifting of the soil."

A method which we have sometimes adopted, and which we think a saving of labor under some circumstances, is as follows:

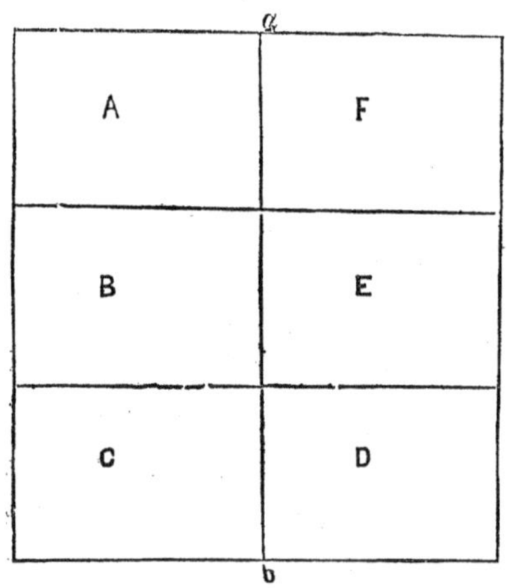

Fig. 1.

Let fig. 1. represent the plot of ground to be trenched. Divide it into two equal parts by the line *a b*, and instead of wheeling the soil out of A F to the rear of the plot, simply throw that from A out in front.

There can, of course, be no more difficulty in finding room for it there than there would be in obtaining a place for it in the rear. Then dig down the bank B, and with it fill the trench A. B is now a trench which may be filled from C; C may be filled from D; D from E; E from F; and the filling of F with the soil which was at first thrown out of A, will make all even. The wheeling of the soil, which is no inconsiderable item, is thus saved. It is evident, however, that this plan is adapted only to small, or at least narrow plots.

All the foregoing operations prove most beneficial when performed in the fall. At that time the soil should not be finely pulverized, but left in as rough a state as possible so as to expose it thoroughly to the action of the winter's frost and snow. It should be also well mixed with a good dressing of well decomposed stable manure, and any of those matters mentioned in Chapter XI.

By these means, the ground will be thoroughly enriched by spring, and will not consist of earth mixed with fermenting masses of manure, than which nothing can be more injurious to young plants. In the following spring the land should be raked or harrowed, so as to obtain a level surface of finely pulverized soil, and if it should be lightly forked over it would be none the worse for it.

TERRACES.—From our directions for the selection of a vineyard site, it will be seen that we prefer a gentle slope to the south or southeast. If this slope does not exceed an angle of eight degrees, or a rise of one foot in seven, it will be unnecessary to adopt any peculiar system of arrangement. For a rise of one in four it will be necessary merely to make very slight terraces, the borders being made eight feet wide and half the descent being taken up by the slope given to them, will leave but twelve inches of a terrace, which may be easily secured by a row of sods, boards or stones, or even the earth beaten hard and kept carefully dressed up. But when the inclination of the ground much exceeds this amount, it becomes necessary to form regular terraces which is best done as follows:

Find out the actual slope or inclination of the ground, which is easily done by taking an eight-foot

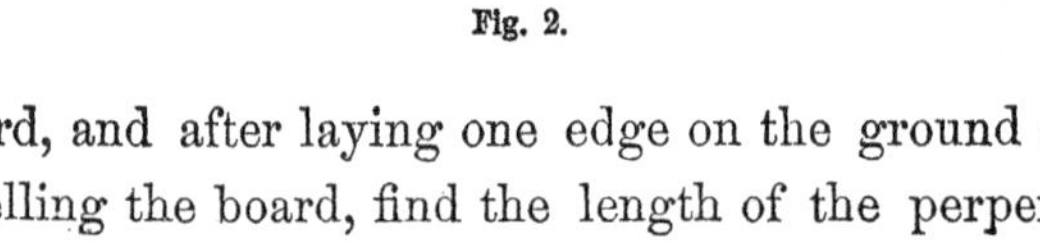

Fig. 2.

board, and after laying one edge on the ground and levelling the board, find the length of the perpendicular which touches the surface beneath the other end. Thus *a d*, fig. 2, being the surface of the hill, and *c* the eight feet board with the level resting upon it, *e d*, will be the rise in eight feet and *e d*, less the slope given to the border will be the height of each step or terrace. Having found this, the next step is to cut a perpendicular face half the height of the proposed terrace at the foot of the hill and against it to build a wall as high as may be required. This is best formed of dry stone, though the bank is sometimes left with a good deal of slope, and sodded, the sods being pinned to the face of the bank with stakes until the roots have penetrated sufficiently to hold. The sods for this purpose should not be cut square, but diamond form, so that the face of the bank would present the appearance shown in fig. 3. But sods are

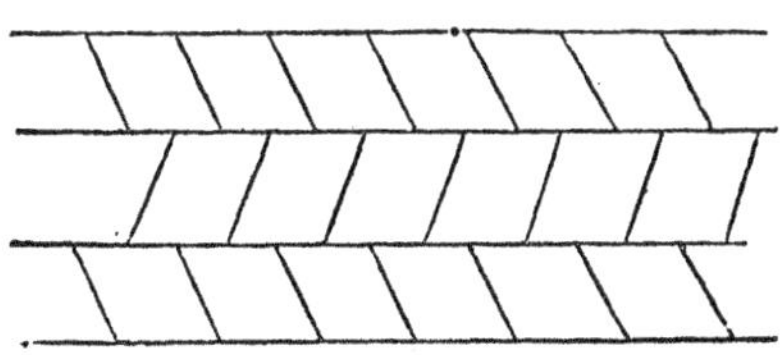

Fig. 3.

objectionable from the fact that they not only keep the air moist in the vicinity of the vines, but also abstract a good deal of nutriment from the soil, and unless kept neatly mown present a very bad appearance. In default of good stone we think that sun-dried brick would make a very good wall. The earth of which they are made should be mixed with straw, well worked and made into blocks.

It is probable that in well-drained terraces such walls would last well if protected with a coping of boards or straw secured with good clay in the manner shown in fig. 4, so as to shed the rain.

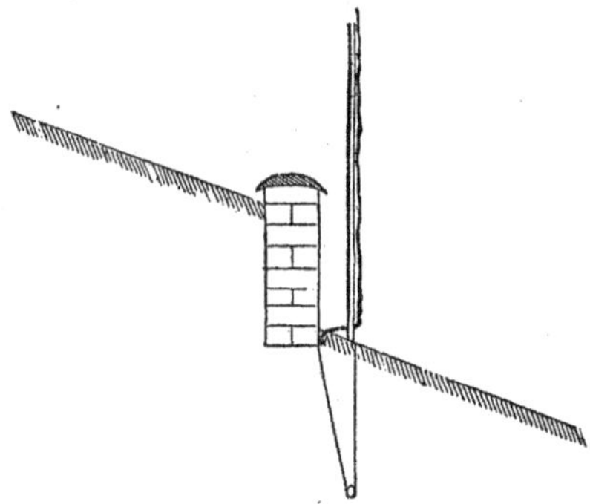

Figure 4.

Having built this wall, the next step is to fill up behind it, and level off a border of suitable width—say 6 or 8 feet. To do this it will be necessary to cut down a perpendicular face the same height as before, when another wall must be built, and the same process repeated.

A writer in the third volume of the "Gardener's Magazine" proposes to train the vines on trellises lying on the surface of the slopes as shown in figure 5. Trained in this manner, grapes are said to have

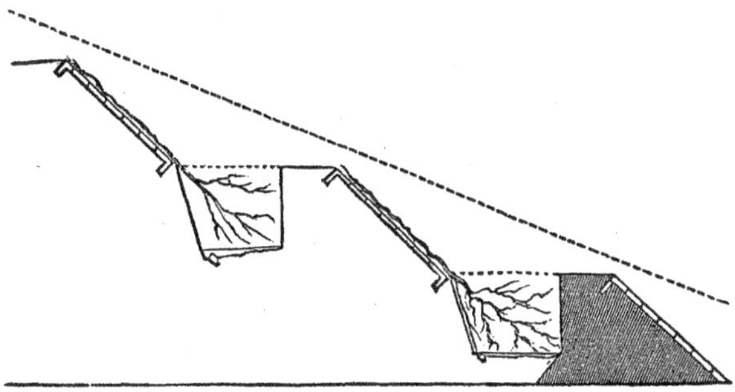

Figure 5.

ripened well in England. We would prefer the vertical trellis, however, and give the illustration, more to show what has been proposed than as an example to be followed. So many times have we seen it proposed to incline trellises and train vines horizontally, that we cannot refrain from quoting Lindley's remarks upon this point.

"That training a tree over the face of a wall will protect the blossoms from cold must be apparent, when we consider the severe effect of excessive evaporation upon the tender parts. A merely low temperature will produce but little comparative injury in a still air, because the more essential parts of the flower are very much guarded by the bracts, calyx and petals, which overlie them, and, moreover, because radiation will be intercepted by the branches themselves, placed one above the other, so that none but the uppermost branches which radiate into space will feel its full effects; but when a cold wind is constantly passing through the branches and among the flowers, the perspiration—against which no sufficient guard is provided by nature—becomes so rapid as to increase the amount of cold considerably, besides abstracting more aqueous matter than a plant can safely part with. To prevent this being one of the great objects of training trees, it is inconceivable how any one should have recommended such devices as those mentioned in the 'Horticultural Transactions,' II. Appendix, p. 8., of training trees upon a horizontal plane; the only effect of which would be to expose a tree as much as possible to the effect of that radiation which it is the very purpose of training to guard against."

All terraces should be well drained, and the drains are best arranged by having a series of cross drains parallel to the terrace, as seen in section fig. 4 and 5, and emptying into a main drain which descends the hill. These drains should be placed as in the figures, taking care to leave the ground under the wall solid and undisturbed. In forming terraces for vine culture it is necessary to exercise care and judgment, so as not to bury the good soil and leave the poor soil for the vines to grow in.

VINE BORDERS.—The formation of vine borders in gardens is a subject upon which the student will find no lack of information, almost every successful gardener attributing the superiority of his grapes to some peculiarity in the construction of his borders, and innumerable have been the paper conflicts waged between the advocates of carrion, asphalte, ventilated borders, etc., etc., and their opponents. The "carrion" controversy has probably caused the shedding of more ink than any of the others, the ultraists on both sides being probably in the wrong. But, after all, we regard the construction of proper vine borders as no very difficult affair, and shall first give our own views in the matter and afterward quote those of other authors.

Of course in borders, as in other cases, it is necessary that the bottom be as dry as possible. This being provided for, if the soil is a light mellow garden mold, we would rest content with trenching it thoroughly, and adding liberal supplies of litter, well decomposed manure, woollen rags, and especially bones;* and if in the bottom of each trench a good

* In the ordinary course of agriculture, where "quick returns," if not "small profits" are an important element of success, bones when used as manure cannot be too thoroughly pulverized. Indeed, it is often profitable to reduce them to the most active form—that of a solution—by means of acids. But for reasons to be hereafter stated, one

layer of brickbats, lime rubbish, and oyster shells be laid, it will prove an advantage. A border prepared in this simple manner will give good satisfaction under any circumstances.

If the soil be heavy we would also make liberal additions of sandy loam or saw-dust.

But if the location of the border is such that it cannot be well drained, we would remove all the soil to the depth of 18 inches over the entire extent of the border and fill up at least 12 inches of the space with stones, brickbats, etc. Over this we would spread a thin layer of straw or brush, and after building a wall round the border 18 to 30 inches high, we would fill in with a rich soil resembling in composition, that described above. The earth on the outside might be banked up to the wall, and either sodded, or merely beaten solid.

In all such cases, it is evident that from the narrow limits to which we are in general confined, the soil ought to be of the richest kind; and as it is nearly impossible to renew it after the vines are once started, this richness should be derived from materials calcu-

great advantage to be derived from the use of bones in vine borders is the length of time during which they continue to act, and, therefore, the largest and most solid should be selected and used without being crushed or broken. This is no argument, however, against the *additional* use of bone dust.

lated to give more than a mere temporary impetus to the plants. The nature and action of manures will form the subject of a future chapter, but we may here state that bones, hair, woollen rags, leather clippings and similar matters are by far the most suitable. For the purpose of giving porosity to the soil, as well as furnishing nutriment to the plants, nothing will be found to equal chopped straw. Chaff, or sawdust comes next in order, and from experiments which we have made on the subject, we do not think the value of the latter is half appreciated. To dead animals, either whole or divided, we have never found any objection, provided they were not placed in direct contact with the roots of the plant. No danger is to be apprehended of the vine seeking them to its own detriment. But this more properly pertains to the subject of manures. We will now give the manner in which the most celebrated grape growers construct their borders.

Miller (1759) recommends good mellow soil without any addition.

Speechly (1790) states in his work: "As the vines in the hot-house at Welbeck have been remarkably fruitful and vigorous, I shall beg leave to recommend the same kind of compost mold which I make use of there, viz. one-fourth part of garden mold, (a strong loam); one-fourth of the sward or turf from a pasture where the soil is a sandy loam; one-fourth of the sweepings and scrapings of pavements and hard roads; one-eighth of rotten cow and stable yard dung mixed; and one-eighth of vegetable mold from reduced and decayed oak leaves. These are the several and respective proportions. The sward should be laid in a heap till the grass roots are in a state of decay, and then turned over and broken with a spade; let it then be put to the other materials, and the whole worked together, till the separate parts become well and uniformly mixed and incorporated.

As the vegetable mold from decayed leaves cannot always be obtained, by reason that the leaves require two years before they become sufficiently putrid and reduced, it therefore may sometimes be necessary to substitute some other ingredient in lieu of this part of the compost; wherefore it may not be inexpedient to point out the proper succedanea.

Rotten wood reduced to a fine mold, such as is often found under fagot stacks; the scraping of the ground in old woods, where the trees grow thick together; mold out of hollow trees, and sawdust reduced to a fine mold, provided it be not from wood of a resinous kind, are in part of a similar nature with vegetable mold from decayed leaves, but are neither so rich nor so powerful, because the vegetable mold receives a power by its fermentation.

Abercrombie directs the top slip of sandy loam from an upland pasture, one-third part; unexhausted brown loam from a garden, one-fourth part; scrapings of roads free from clay, one-sixth part; vegetable mold or old tan, or rotten stable dung, one-eighth part; shell marl, or mild lime, one-twelfth part. His borders he recommends to be from three to five feet in depth, and where practicable, not less than four feet wide within the house, and not less than ten feet wide without.

The vine borders at Wishaw House, Lanarkshire, in a cold and wet locality, are thus formed: Breadth, 12 feet, depth of soil 18 inches, under which is laid a foot of hard clinkers, by way of drainage. The soil used is that natural to the garden, which had for years been under pasture, and is a remarkably strong, rich brick-clayey loam, with no other preparation than the addition of a moderate supply of stable manure. In this soil the best grapes ever produced in Scotland have been grown for the last three years.

A writer in the "Gardener's Chronicle" (1843, page 825) prepares his borders thus: The soil most suitable for a vine border is the surface spit from a field of an old fertile loam pasture; this should be collected some time before it is required, mixed with a good proportion of cow dung, and the whole turned over at intervals, three or four times, and exposed to the action of the weather. In preparing the border, the old earth should be cleared away from the whole space, to the depth of about two and a half feet, and a main drain cut parallel with the length of the border, at its extreme outer edge.

This should be at least two feet lower than the bottom of the border, whether laid with concrete, chalk or bricks, and the bottom of the border should have a gentle inclination from the back to the drain. To render this drainage more effectual, cut small drains, placing drain tiles at their bottoms, at convenient distances, to run in a slanting direction from the back of the border into the main drain, the latter being six inches below them. A few turfs should be laid over the tile drains with the grassy side down; the fresh soil may then be filled in, taking care to keep the roughest part near the bottom.

Three cubic yards of compost are enough for each vine; this will admit of the border being ten feet wide, or with forty-eight cubic feet, you may form it only six feet wide in the first instance, and add six feet more as the vines extend.

Roberts, the great advocate for carrion, gives the following description of his border: "The compost and manures I most recommend, and which I made use of, are two parts the parings of a piece of old pasture land, a strong loam laid up one year (or till the sward is half decomposed), in the form of a potato hod, close covered in with soil, and never turned; one part, the turf with four inches of the soil, of a looser texture laid up for the same period, and not turned, as before; an eighth part scrapings of the highways formed from limestone, or other hard material; and the other eighth part, half decomposed horse or cow dung. I am not an advocate for turning over and mixing the materials promiscuously together, as, by often turning, the compost becomes too solid, losing a great portion of its fertilizing property by such repeated intermixture; and unless it be of a very sandy, loose texture, the border will, in a few years, become impervious both to water and to atmospheric air, which are of incalculable benefit to the growth of the vine. I would recommend the autumn, if the weather be dry, to prepare to fill in your border.

"A month previous to filling your border, provide a quantity of carrion, cattle dying by accident, disease, etc., which I am sorry to say, has, of late years, been too common an occurrence. If you have collected it some time before hand, have it cut into small pieces and laid up in soil till the time of using. It emits a very nauseous effluvia, but this must be borne, for this is the *pabulum* to produce the nectar of Bacchus. When all is ready, and the weather favorable, proceed at one end of your border, wheeling in and mixing the materials in proportion as they stand to each other in my previous directions, on no account breaking the materials in mixing, but turn them in as rough as possible, adding one good sized horse or cow carcass to every ten or twelve square yards, using caution, and not bringing it to the surface of the border within one foot as its assistance is not wanted the first year. What I have here recommended is my practice adopted at this place, the result of which, I dare presume to say, has surprised all, both gentlemen and practical gardeners, who have witnessed it."

Fiske Allen, one of the best American culturists of the vine under glass, constructs his borders thus:

"If the soil is very poor, or unsuitable for the purpose, so as to require to be removed entirely, then a compost prepared thus is recommended; one-half to be the top soil of an old pasture, one-quarter to be bone, or some other strong manure; one-eighth oyster shells, or lime and brick rubbish; one-eighth rotten manure; these articles thrown together in a heap, and so to remain until decomposed and amalgamated, when they should be placed in the border and thrown loosely together. My borders having the most slaughter-house manure, or whole bones of animals in their composition still continue, as they ever have done, to produce the best fruit and the largest crops.

"It is unnecessary to attempt to give rules for every kind of soil. One must use his own judgment, and make his border to consist, as near as can be, of the above ingredients. He must bear in mind that, if his soil is a stiff clayey loam, he must add freely of such materials as will lighten and give permeability to it. If the soil is light, sandy or gravelly, with the manure should be added a proportion of clay or clayey loam. The rich alluvion soil, abounding in our western and southwestern States, will not require any of these strong manures. If anything is requisite to improve them, it must be shells, charcoal, leaves, small stones, or gravel—such materials as will loosen the soil."

But that the reader may not be discouraged by these extravagant demands we quote the following from Hoare:

"But if vines could not be planted with any prospect of success in any other situations than in borders set apart for that purpose, but a very small quantity of grapes could be grown, compared with what the country is capable of producing. Innumerable instances occur throughout the country, and especially in towns and their suburban districts, in which walls, cottages, houses, and various descriptions of brick and stone erections present very favorable aspects for the training of vines, but which nevertheless are so situated locally, as to possess little or no soil at all on the surface adjoining their sites; the ground being either paved with bricks or stone, or perhaps trodden so hard, as to be apparently incapable of yielding sustenance to any vegetable production.

"In all such cases, however, if the ground adjoining the site of the wall or building be opened to the extent of eighteen inches square, and as many deep, it will be sufficient to admit the roots of a young vine, which must be pruned to suit that space. If a wider and deeper space can be made, it will of course be better; but if not, that will do. After the sides and bottom have been loosened as much as possible, the vine may be planted and the hole filled up with two-thirds of rich loamy earth, and one-third of road scrapings, previously mixed well together, and if necessary the surface covering, whether of stone, brick, or otherwise, may be restored again to its former state, provided a space about six inches square be left open for the stem to swell in during its future growth. Vines planted in such situations, will in general do well, although their growth will not be so rapid as when planted under more favorable circumstances.

"In all cases where vines are planted against any description of buildings, their roots push as soon as possible under the foundations, being attracted thither by the warm air which is there generated; and such situations being also dry, from the excavations which have been made, offer to the roots the same protection from excessive moisture, as the substratum of a well-prepared border. The same may be observed of vines planted against walls, the foundations of which possess similar advantages, although in a more limited degree. Hence the fact may be inferred that vines planted in such situations, without any previous preparation of the soil, will frequently grow as luxuriantly, and produce as fine grapes as those planted in rich and well-prepared borders.

"Indeed, it is hardly possible to plant a vine in any situation in which it will not thrive, provided its roots can by any means push themselves into a dry place, and the aspect be such as to afford to its branches a sufficient portion of the sun's rays to elaborate the juices of the plant.

"The truth is, that the roots of the vine possess an extraordinary power of adapting themselves to any situation in which they may be planted, provided it be a dry one.

"They will ramble in every direction in search of food, and extract nourishment from sources apparently the most barren. In short, they are the best caterers that can possibly be imagined, for they will grow, and even thrive luxuriantly, where almost every description of plant or tree would inevitably starve."

CHAPTER IV.

PLANTING.

Proper Age of Vines for Planting.—Where young vines have been raised from cuttings, in the open ground, two years old probably is the best age to select for planting out. Plants one year from the cutting have rarely made sufficient roots to bear transplanting well, and at a greater age than two years the roots are so long that they generally receive much mutilation in taking up—thus losing their most fibrous and valuable part, viz., that at the extremities. Of course older vines, carefully taken up and as carefully planted, will come into bearing in shorter time than younger plants, and thus give more satisfactory results where expense is no objection. But where a large number of vines are to be set out, two-year old plants, as above stated, or one-year old plants raised from eyes in the spring, and grown all summer in the open air, have decided advantages on the score both of economy and ease of planting. Indeed, we should prefer plain cuttings, planted two to each stake, to one-year old vines raised from cuttings in the open ground. Plants raised from eyes in pots, early in spring, and transferred in summer to their final location, do very well.

PROPER SEASON FOR PLANTING.—The proper season for planting depends much upon local circumstances—soil and climate being chiefly to be considered. In a few instances, were the soil is light and the climate mild, it may do to risk fall planting, but under all ordinary circumstances we should advise this operation to be deferred till pretty late in the spring, and this advice is founded upon the uniformly favorable results which have attended this plan in our own experience, as opposed to frequent want of success at other times. Plants set out even early in the fall rarely outstrip those planted in the following spring, and when autumn planting is delayed much beyond the fall of the leaf, the plants frequently fail if the winters are severe.

The reason of this probably depends upon the fact that the roots of all plants when vegetation is active, are enabled to resist adverse influences which would prove fatal to them when dormant. Thus the vine when growing will revel in a degree of moisture which would destroy it, or at least prove very injurious during the winter months. Now the roots of all trees are more or less injured by transplanting, and incipient decay is apt to supervene unless the vitality of the plant is sufficient to withstand it. If this should occur when the plant is dormant, there is no influence at work to resist the evil. But if such injuries should be inflicted in spring, when vegetation is just commencing, they are quickly and readily repaired.

With care vines may be transplanted even when their leaves are well developed; but under such circumstances the vine, from its great evaporative powers, makes a heavy draft upon the roots and is rather impatient of removal after vegetation has made some progress. We have had the best success, however, when the plant was set out so late in the spring that the buds were starting, but just before they were fully burst. About this time the soil is getting gradually warmer, and although it does not reach a sufficiently high temperature to induce the formation of roots in cuttings before June or July, still it is warm enough to allow of the healthy action of the roots in a growing plant.

In the above cases the vines were set out immediately after being dug up. Where it is necessary to transport them any distance, it would undoubtedly be better to take them up earlier, before the sap begins to move. They need not be planted for some time, but may be merely *heeled in*, as it is called, that is, placed in a shallow trench and well covered with dry soil. A covering of straw or leaves in addition will do no harm if the mice do not make it a harbor.

Distance Apart.—The distance apart at which vines should be planted will, of course, depend not only upon the variety, but upon the object for which they are set out. In Europe they are placed at all distances from 30 inches to 30 feet. In the Ohio vineyards, where they are usually fastened to stakes, the plants are placed about four or five feet apart; but in the northern States, where vines are trained upon trellises, we should prefer to set them out in rows 6 feet apart, and the vines standing 7 or 8 feet apart in the rows. This distance enables us to keep the vines close enough and short enough for all practical purposes, while it does not require more time to cover the trellis than is absolutely necessary to bring the vine into proper order for bearing. The rows are also sufficiently far apart to allow of horse labor being used—a considerable saving being thus effected.

The number of vines required to plant an acre will be seen from the following table, which has often been published, but which it may be well to insert here.

		PLANTS TO THE ACRE.			PLANTS TO THE ACRE.
3 × 3	requires	4,840	8 × 7	requires	788
4 × 3	"	3,630	8 × 8	"	680
4 × 4	"	2,722	9 × 7	"	691
5 × 4	"	2,178	9 × 8	"	605
5 × 5	"	1,742	9 × 9	"	537
6 × 5	"	1,452	10 × 7	"	622
6 × 6	"	1,210	10 × 8	"	544
7 × 6	"	1,023	10 × 9	"	484
7 × 7	"	888	10 × 10	"	435

MARKING OFF THE GROUND.—Where vines are set out at from 4 to 6 feet apart and trained to stakes, the following directions, taken from the "American Philosophical Transactions," and frequently quoted (generally without credit), are as good as any:

"Your squares being laid out, and having concluded how far your vines shall stand every way from one another, in which every man is to please himself, you stretch a line of proper length, and stitch small pieces of red, blue, green, or any other colored cloth, at such distance from each other as you mean to plant the vines. I will suppose 8 feet, because upon the most mature deliberation, I think that the best distance for vines to stand in this country, as I shall afterward show more fully. The line being ready, stretch it along the head or upper part of your square, so that a rag appears at each corner, drive down a stake at every rag. This done, move your line down to the lower side of the square, which is opposite to the first, and stretch your line along that, having a rag at each corner, and drive down a stake at every rag. Then turn your line the other way up and down, and fasten your line to the upper and lower outside stakes, so that a rag be at each stake, and drive down a stake at every rag, and so on from stake to stake, till the whole be completed. If you have been careful not to disturb or move the line, when you drove down the stakes, and have driven them all on the same side of the line, your square will be uniform, and the stakes near the ground will range exactly every way."

Where the vines are trained to trellises, it is not essential that they should be straight both ways as when tied to stakes, it being necessary that the trellises be parallel and equi-distant only.

To make them so, a very good plan is as follows: Prepare a rod, fig. 6, a few inches more than twenty feet long, and having a small hole (*a*) bored through one end, then bore a similar hole (*c*) twelve feet from the first; one (*b*), 16 feet from the first; and one (*d*) 20 feet from the first. Having decided upon the direction of the first row and divided it into spaces corresponding to the distance the plants are to stand from each other, drive in stakes A A at each end, and measuring off 16 feet, drive in other two, B B. The heads of these should be made level with

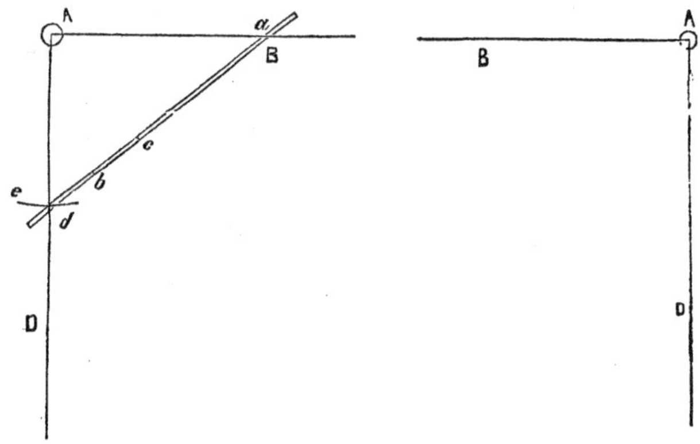

Fig. 6.

the surface of the ground, and headless nails should then be driven into them—the two nails (A and B) at each end being exactly 16 feet apart. Now place the rod on one of the outside stakes, so that the nail will pass through the first hole (*a*) and drive a peg into the hole in the rod 12 feet from the end. With this peg mark a curve (*e*), and then placing the end (*a*) of the rod on the stake B, mark another curve (*d*) crossing the first by means of a pin passing through the hole (*d*). A line (A D), drawn from A through the crossing of these curves will be perfectly square with the first row. Divide the lines (A D) into spaces equal to the width of the rows, and the lines joining the corresponding divisions will be the proper lines of the trellises.

As many stakes or poles, 6 or 8 feet long, should be provided as there are vines, and these should be set at the points where the vines are to be placed before the holes are dug. These stakes will serve during the first two years and will save the tear and wear of trellises as well as the interest on the investment, besides relieving some of the hurry incident to the work of the first spring, which is always more pressing than that of any succeeding one.

Digging the Holes.—In digging the holes it will be well to take up the stakes one at a time, and after digging the hole to replace the stake, driving it slightly into the soil at the bottom. By so doing the centres of the holes, or at least the points where the stems of the vines should come, will be easily kept, whereas if all the stakes were removed before digging the holes, it would be troublesome to get them exactly right again. The insertion of the stake, before planting the vine, not only aids us in this, but prevents the possibility of injuring the roots by driving a sharp stake through them, a thing which is often done. The holes should be dug from 12 to 24 inches deep and about 3 feet in diameter, or as wide as the distance to which the roots extend.

Taking up the Plants.—When plants are purchased, this operation is generally left to the nurseryman who not unfrequently commits it to men who care very little *how* it is done so that they get the plants out of the ground. But when we reflect that a small amount of care in taking up a plant will often cause a difference of a year or more in its subsequent growth, it will be evident that the labor, time and consequent interest on capital which will be ultimately saved by devoting a little attention to this matter will more than pay for the few extra minutes required.

In taking up a vine, or any plant, it is well to remember that the most efficient portion of the roots is that which lies at the extreme ends—those minute fibres or spongioles which have been aptly termed the mouths of the plant. In old vines, where the roots extend to a great distance, these fibres are generally left in the soil, and the plant presents but a few smooth, fibreless, cord-like roots from which spongioles must be emitted ere the plant can derive any nourishment from them.

But in young plants the roots have not yet extended so far as to prevent their being easily taken up without any great loss, and here we have one reason why we prefer young plants to old ones. A consideration of this fact will also lead us to follow out each root to its termination and so secure all the fibres possible. Where the plants have been started in sandy, friable soil this is not a laborious task, but where the soil is clayey and plastic, it is often a work of considerable difficulty.

In any case, however, the purchaser should see that the roots are taken up as completely and entirely as possible, and it will also be well to prune all that are bruised, broken, or diseased—taking such injured parts off with a clean cut. As these injuries can never be so well seen, or so well remedied as when the plant is newly taken up, this is the best time to attend to them; and this forms another reason why the purchaser should, if possible, give this matter his personal supervision. They should then be dipped in puddle made of good garden soil, stirred up with water; clay is frequently used for this purpose, and so is cow dung, but very injudiciously. The former is too tough and hard and prevents the formation of young roots, while the latter (as it is commonly used) is caustic and destructive to the tender fibres. We have tried all three substances and are confident that nothing will be found equal to good common soil.

If the nursery in which the plants have been raised

is on the same premises as the vineyard, the vines may be either rolled up in coarse sacking, or, a few being taken up at a time, they may be plunged in a pail or tub filled with puddle. But if they are to be sent to a distance, they should be packed in damp (not moist) moss (sphagnum) or good clean straw, and either made into light bundles or firmly packed in boxes.

The vines having been carefully taken up and the holes properly prepared to receive them, the next step is to set them out, and in doing this the following points require special attention:

1. That the roots be disposed in their new location as nearly as possible in the same position that they occupied before their removal.

2. That some fine, friable, mellow mold be placed in immediate contact with the roots.

3. That no fresh manure or decomposing organic matter be allowed to come in direct contact with the plant under any circumstances whatever.

4. That the soil be firmly packed about the roots, no air spaces being left. In doing this, however, do not tread down the plant with your whole weight, as you will thus be very apt to tear off some of the roots, but work the soil in with your hand or a pointed stick.

In general it will be well to insert the plants a little deeper in the soil than they were previously, as, owing to the mutilation which of necessity takes place, a greater draught is made upon the roots for moisture than they can support when thus shortened, if they are placed near the surface. But this point requires the exercise of discretion, and a good substitute for deeper planting will be thorough mulching.

As roots always spring from a bud or joint, and rarely from the internode or portion between the buds, the mode of propagation by which the plant has been produced will exert considerable influence upon the *modus operandi* of setting it out.

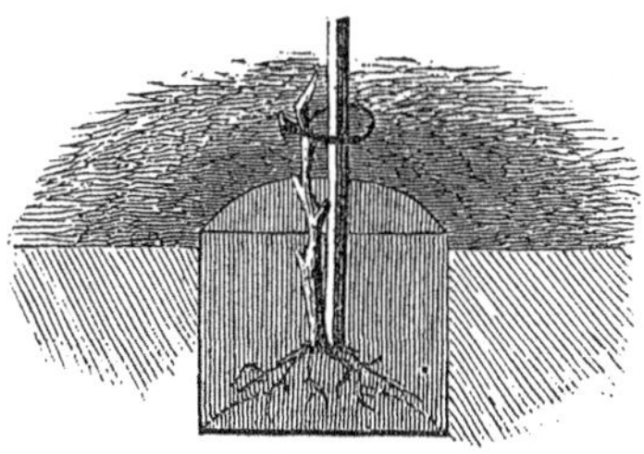

Fig. 7.

In fig. 7 is shown the proper disposition of the roots of a young plant raised from a seed or from a single eye. In this case the roots all proceed from within a short distance of each other and from the base of the stem. In setting out such a plant, the better plan is to throw a shovelful or two of fine mold on the bottom of the hole, so as to form a conical heap, the top of which should be just at a suitable height to support the base of the stem in its proper position. Then, having placed the plant on the top of this little mound, spread out all the roots equally and naturally over its side and fill in with pulverized soil, being careful to pack the soil firmly around the roots, yet still leaving it mellow and porous.

The soil ought to be raised some inches above the surrounding ground, the amount depending upon the size and depth of the hole dug. All filling-in is apt to sink, and unless this is done, the plant may be found after a few weeks to be too low. Some, however, prefer to have the plant set in a hollow, claiming that a basin is thus formed which catches and retains the rain. We would rather rely upon good mulching for obtaining the requisite amount of moisture, but if this is dispensed with, and recourse had to the former plan, we should prefer to have the hollow or basin in the form of a ring around the edge of the hole, leaving the stem surrounded with a little mound which will shed the rain. The stem is thus kept dry, and the moisture is guided just where it is wanted, viz.: to the extremity of the roots. Figure 8 gives a sectional view of the soil so arranged.

When vines have been raised from cuttings con-

Fig. 8.

sisting of several eyes or joints, there will in general be several layers of roots—the plants having the appearance shown in Fig. 9.

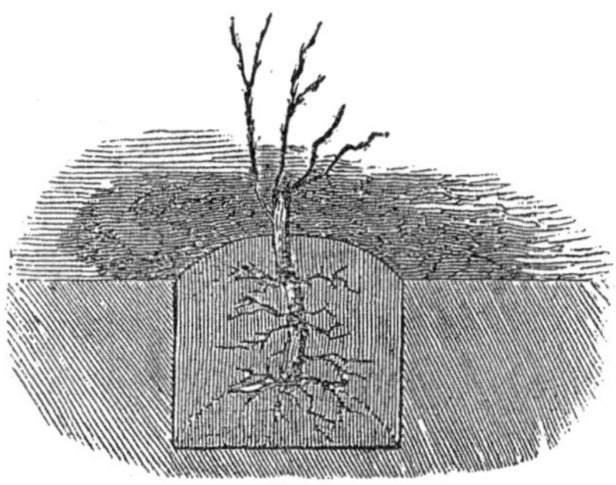

Fig. 9.

In this case it will be necessary to proceed as directed for plants from eyes in so far as the lower layer of roots is concerned—the upper layers being held up while this is done. After the first layer has been pròperly covered, the next layer is laid on the surface of the soil which covers the lower layer, and after being properly disposed are covered in turn, which process is repeated until all the roots are imbedded in the soil.

Plants raised from layers in general demand a treatment peculiar to themselves. If they have been produced as in Fig. 44, page 176, they will, of course, be set out in the main as shown in Fig. 7. But if they have been raised in the open ground, and the roots have been produced from several joints or buds, it will be found that while the roots are not disposed in regularly-ascending layers, yet that some are lower than others—the whole, however, in general lying in one plane which is greatly inclined to the surface of the earth. For such plants it will in general be best to dig a trench or oblong hole, and instead of raising a heap in the centre to lay the soil in the bottom, so as to form a regularly inclined bed. The plant being placed on this bed of fine soil, the roots are all arranged over it at once and covered in without further trouble.

In all these cases it will, of course, be necessary that the stem of the plant be placed sufficiently near to the stake which has been inserted in the hole to allow of its being tied thereto without much bending or wrenching, and if the weather be dry it will be necessary to give the plants a good watering at the time they are set out.

When plants are received in pots—having been

grown therein from eyes or grafts—it is always best after taking them out to remove a considerable portion of the soil, and spread out the roots. This is necessary from the fact that the roots of plants grown in pots form a series of spirals round the outside of the ball (between the earth and the pot), and if set out in the ground just as they are taken from the pot, it requires a long time before the roots change this habit and acquire a proper direction and healthy condition. The plants should be well watered before being taken from the pots, and they should be set in fine, loose soil, being exposed to the air as little as possible. After planting, it will, of course, be well to be liberal with water, and liquid manure used in a very diluted state will prove highly beneficial *after the first week.*

Plants for setting out are usually obtained in pots in June, July, or the beginning of August, and as it frequently happens that at that time the earth and air are so dry as to endanger the life of a young vine, if treated as just directed, we have sometimes found it advisable after receiving them from the nursery to set out the pots (without removing the plants) in the open ground, plunging them about two inches below the surface of the soil, and leaving them there until a few rainy days occur, when the pots are taken up and the plants removed and properly set out.

CHAPTER V.

CARE OF THE VINES DURING THE FIRST, SECOND AND THIRD YEARS.

The roots of the vine having been properly cared for, the branches may now be pruned. Unless where very large and well-rooted vines have been planted expressly for immediate bearing, all the secondary shoots should be cut away and the main stem shortened to an extent depending upon its character.

As usually received from the nursery, one or two year old plants, if raised from cuttings, consist of a short stem two to six inches long, one or two shoots and a large quantity of spray or small twigs, consisting of the laterals of last year. If raised from eyes, there will in general be but one shoot, with perhaps a few laterals. Under any, circumstances the plants ought to be cut back at planting to two good eyes, and as soon as they have made a few leaves, cut off the upper one as close as possible to the one left, taking care, however, not to injure the base of the remaining shoot, which ought to be kept tied up to the stake as fast as it shows symptoms of leaning over.* The

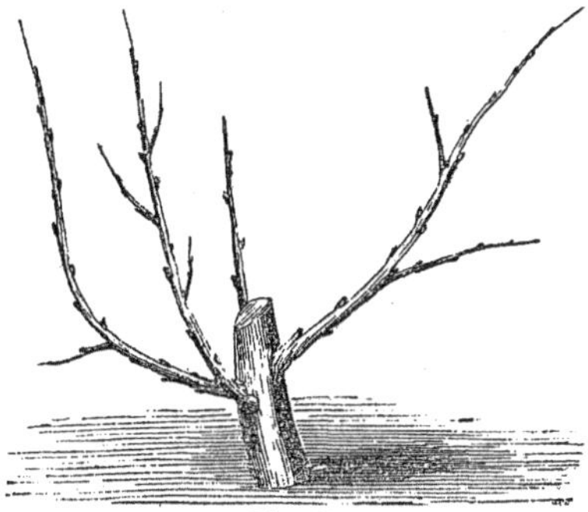

Fig. 10.

base of the shoot which is retained (that is, the point

* It is recommended by some respectable authorities, to allow the young plants to remain untied during the first year, urging as a reason that more vigorous stems will thus be obtained. But, although the experiments of Knight have proved the advantage of bending and motion to most young trees, yet the vine naturally seeks support from surrounding objects, and will in most cases, receive more injury from dirt, and abrasion by being blown about and rubbing upon the ground than will balance the good derived from the motion imparted.

But as we may observe that the vine is adapted to cling, not to the thick and stout bodies of trees, but to slender branches, it is obvious that nature provides fully for sustaining the plant beyond the reach of injury, without interfering with the action of the wind in producing motion. Hence, in the construction of trellises and the choice of stakes, it will be well to select flexible material, always provided it is strong enough to avoid all danger of being blown down ; stout rods or poles are therefore to be preferred to sawed lumber, and we may add they are also cheaper. From the above facts we may also gather the reason why wires are to be preferred to wooden slats in the construction of trellises.

at which it springs from the old wood) should be as low down as possible—if even with the surface of the ground, so much the better.

A plant such as we have described, is shown in Fig. 10 as it is usually received from the nursery. The same plant properly pruned is shown in Fig. 11.

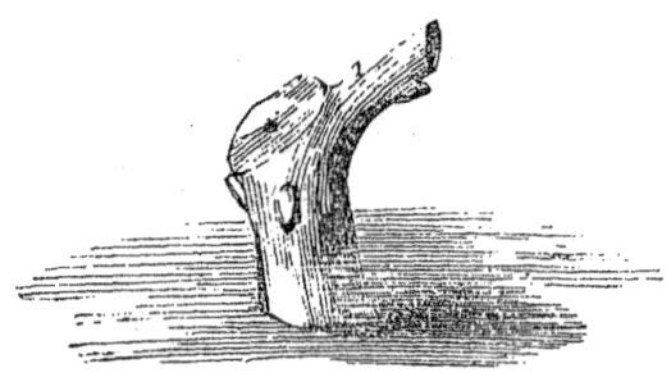

Fig. 11.

Many are afraid to cut back so severely, but it is the only true method.

The object of leaving two eyes at first, is merely to guard against accidents. If we could be insured against them, the upper one would be better away. Little else can be done during the first year than to keep the ground mellow, loose about the plants and free from weeds. The vine must also be tied up during the season, and if a little liquid manure could be applied to them while growing, it would prove of great benefit. In applying this stimulant, it is necessary to use it in a very diluted state, and if possible, just before or during wet weather. When applied during very dry weather remove the surface soil to a depth of three or four inches, and give at least a pailful to each plant, working the soil as little as possible, lest it be converted into puddle. Such an application will last for ten days during even very dry weather, and will do more good than frequent sprinkling.

MULCHING.—But if abundance of grassy weeds, litter, stable manure, or similar matters can be obtained, the best plan is to mulch the plants deeply for at least three feet every way from the stem. Of this process, A. J. Downing says: "Covering the soil in summer is, in this country, one of the most valuable aids to good cultivation ever put in practice. The best mode of doing this is, by what is technically called *mulching*. This consists in spreading over the surface of the ground, so far as the roots of the tree or plant extend beneath it, a layer of tan bark, saw dust, barn yard litter, straw, salt hay, sea weed, or the like, of sufficient thickness to maintain, as nearly as possible, a uniform state of temperature and moisture for the roots. From an experience of some years, we do not hesitate to say that mulching the surface of the ground over newly-planted trees, is not only far better than any after-watering, but that, if the layer is thick enough to keep the surface cool, it renders water wholly unnecessary. In the case of bearing fruit-trees, especially the more delicate kinds, as dwarf pears, apricots, etc., mulching not only precludes the necessity of stirring the soil, by preventing weeds from growing, but it conduces so much more to the health of the tree, and the size and excellence of the fruit than any other practice in horticulture, that the more intelligent growers in the United States now consider it indispensable in this climate."

In addition to these lucid directions, we would only say that before applying mulch of any kind to a young vine it will always be advisable to raise the soil around the stem to the depth to which it is intended to lay the mulch, so as to prevent any of the latter from coming in contact with the plant, as in this case it might be productive of evil.

The annexed figure, 12, where the mulch is seen on the surface of the soil, will illustrate our meaning.

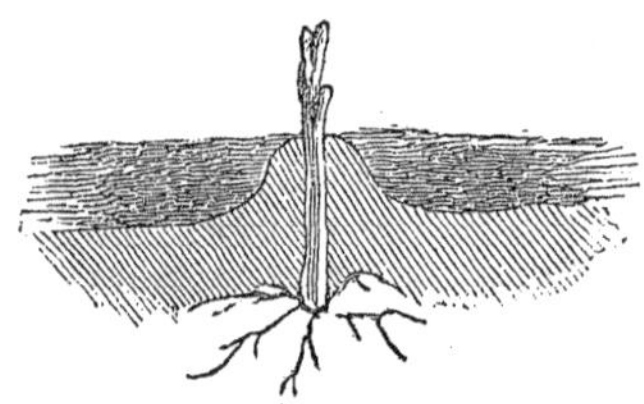

Fig. 12.

LATERALS* are small shoots which spring from the axils of the leaves (the point which they join the

* To these the French have given the name *stipulaires*, and it seems to us that *stipularies* would be quite as good a term as laterals, and more correct.

shoot). As these laterals absorb much of the nutriment which would otherwise go to the increase of the stem, they should be carefully pinched out after they have made one or two leaves. If removed before they have made some growth, the bud at their base is very apt to *push*, as it is called (that is, to grow), which should be avoided, if possible.

Fig. 13 shows a young shoot of the current year with a lateral (B) springing from the base of the leaf L. This lateral should be pinched off at the cross line.

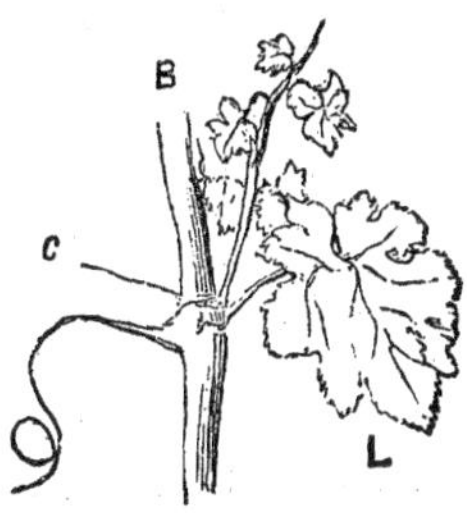

Fig. 13.

If removed entirely or too soon the bud (C) will be apt to push, and destroy our prospects for next season.

WINTER PROTECTION OF YOUNG VINES.—At the close of the season, the vines may either be bent down and covered with earth in the manner usually adopted for covering raspberries, or they may be left upright, and tied to the stakes, a mound of earth being raised up around each such mound, being at least 18 inches high. The soil of which it is made should be taken from the centre of the rows, as, if we take it from about the plants, we only cover the stem to expose the roots.

Where the vines are left tied to the stakes, we prefer to leaving them unpruned. True, most of the wood gets killed, but this is of little moment since it is to be nearly all cut away at the spring pruning.

MANAGEMENT DURING THE SECOND YEAR.—As soon as the severe frosts of winter and early spring have passed away, uncover the young vines, and if not already pruned, cut them to a good bud within 9 to 14 inches of the ground. They should be shaded for a few days from the sun and cold, which may be very well done by sticking a shingle before each, though two shingles placed so as to form an angle in which the vine may stand, will be better. We have now arrived at a point where it will be necessary to decide upon the peculiar system to be adopted in the training of our vines. Instead, however, of describing all the different modes of pruning and training in this place, we shall give only that which we consider best adapted to the native American varieties and leave the consideration of the others to the chapter on general pruning and training.

If the plants have made but a weakly, stunted growth, it will be necessary to allow them another year before proceeding to grow shoots for permanent arms or branches. In this case but one shoot should be trained up, which may be treated precisely as directed for the first year. But if a cane of from 6 to 12 feet has been produced, we may safely proceed to train up two canes which will serve for the future arms of our vine. To do this, after cutting down the first year's shoot as directed, remove all the buds except the three uppermost, and as soon as these are beyond danger of accident, rub off one if three

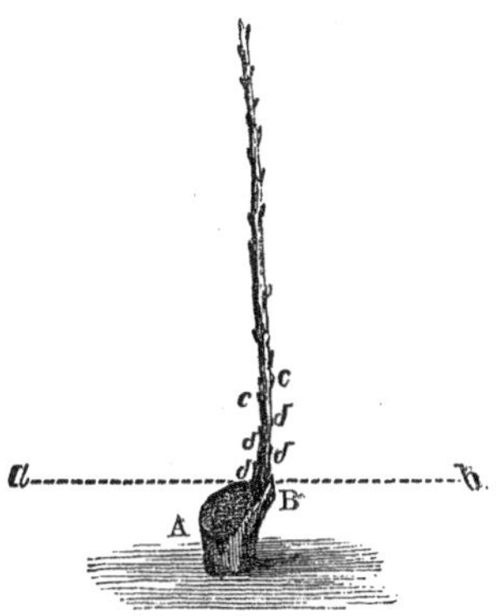

Fig. 14.

should still remain. The two shoots which are left must be carefully trained up, the laterals being pinched out and any fruit blossoms which may appear being removed.

The operations of the second year will be readily understood from the inspection of Fig. 14, which shows the young vine as it should appear at the close of the first season. Here *a* is the wood of the old cutting, with *b*, the shoot which was cut back and from which the young cane *c d* grew. As this old wood is hard and cross-grained and *cannot* be renewed, it will be well to add top dressing, sufficient to cover it up to the line *a b*. The buds, *c c*, are those which produce next year's shoots; and the buds at *d d* must be carefully removed.

The ground should be kept clean and mellow during the season, and by the first or middle of September the further growth of the canes should be stopped by pinching off the ends—the wood being much more thoroughly ripened when this is done.

It will be necessary, or at least advisable, to lay the vines down this season also and protect, not only the old stem, but at least four feet of the young shoots. The stakes may be removed, and during the fall or early spring the trellises may be erected, for which full directions are given in Chapter IX.

MANAGEMENT DURING THE THIRD SEASON.—The trelises having been constructed in such a manner, that the lowest slat or wire may be just below the base of the second year's shoots, that is from 9 to 14 inches above the surface of the ground, these two shoots should be firmly, though not tightly, tied, in a horizontal position as shown in Fig. 15, and all buds

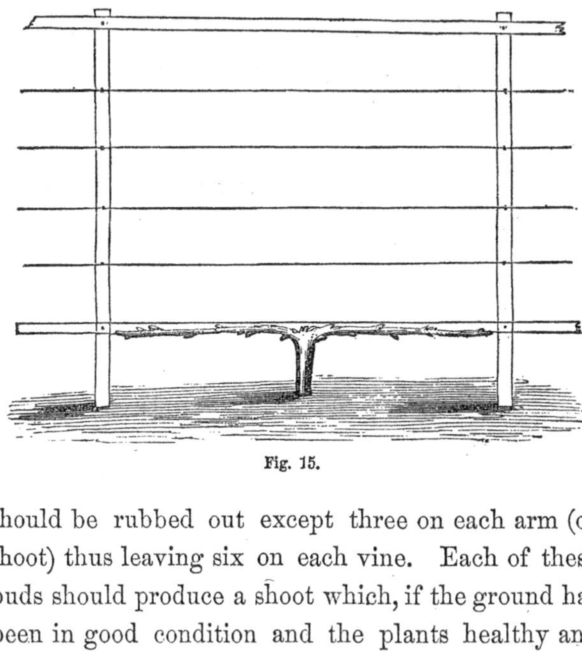

Fig. 15.

should be rubbed out except three on each arm (or shoot) thus leaving six on each vine. Each of these buds should produce a shoot which, if the ground has been in good condition and the plants healthy and properly set out, would reach from 12 to 25 feet unless stopped, and as it is upon every second one of these that we depend for our next year's supply of fruit, they deserve and will require great care and attention in order that they may finally be of equal strength and well ripened. Every second shoot should be stopped when it has made a growth of about two feet, and if any of the others should so far outstrip their compeers as to reach the top of the trellis much before them, they should be stopped also, though except in the case of excessive growth all the shoots had best be allowed to grow on until the first of September, when they may all be stopped at once, unless it be deemed best to allow the weakest a few days' longer growth, in which case it is surprising how soon they will overtake their companions.

Stopping, or pinching, consists in breaking off the end of a shoot, and its immediate effect is to arrest the further growth of the cane, or at least its further lineal development, for the time being. But although no more leaves are immediately formed, those already in existence perform their usual functions and the whole energies of the plant are directed to the ripening of the wood already produced. After a time, one of the buds near the extremity of the shoot will probably break and become the leader, when it should be stopped in turn, this process being repeated as often as any symptoms of vigorous growth are exhibited. The result of all this checking is to lessen the ultimate amount of wood produced and to improve its quality both as to ripeness and density.

Stopping furnishes us with an effectual means of

equalizing the growth of our young canes—a most important point, not only as regards the neatness of their appearance, but the regularity with which the fruit buds will break next season and the strength with which they will shoot. But as the latter points depends not only upon the *size* of the canes, but their *maturity*, it is necessary that an equal growth be kept up during the whole season. This is easily accomplished as the stopping may fortunately be performed at any time.

The same directions as to the removal of laterals and the clearing of the ground should be observed during this as during former years. Greater care is, however, required in the treatment of laterals when raising fruit-bearing canes, as if by too close pinching we should cause the buds which are found at the base of the leaves and upon which we depend for our next year's fruit to push, our prospects would be materially injured. A good rule will be, never to pinch out the laterals, and stop the main cane at the same time; and if the vines show a very vigorous growth of wood, to allow the laterals to make two leaves before stopping them. If the vines are weakly, we may stop the laterals as soon as they appear, as in this case, the main shoot makes sufficient draft upon the roots, to keep all other growth in abeyance.

CHAPTER VI.

MANAGEMENT OF FRUITING VINES.

At the close of the third season we ought to have a vine such as is shown in Fig. 16, consisting of a stout, strait, clean stem, 9 to 14 inches high, from the

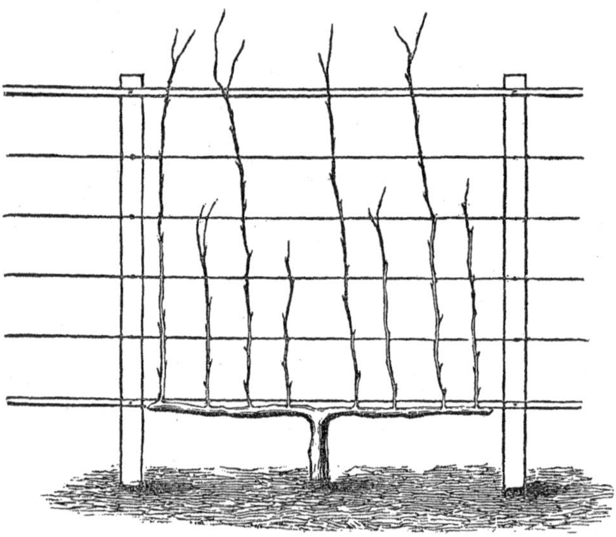

Fig. 16.

top, or head of which springs two horizontal arms, each bearing two well ripened canes, 8 to 10 feet long, and two smaller shoots of from two to five feet. The two canes ought next season to produce 3 to 5 lbs. of fruit each, and their proper care during the winter is worthy of our best efforts.

WINTER PROTECTION OF THE FRUITING CANES.—As the vines have now assumed their permanent form and size (unless it should be deemed advisable after the lapse of a few years to remove each alternate vine and so double the extent of trellis allotted to the remainder), it becomes important to settle upon a systematic course of procedure in order to facilitate our operations, and this remark applies to their protection during winter as well as to every other process connected with them. Of the advantage, we had almost said necessity, for winter protection there can be no doubt. Some extensive cultivators, at a late meeting of the Western N. Y. Fruit Growers Society stated, that they would have made $100 per day for the time spent in covering their vines if they had done so in the fall of 1858.

One gentleman asserted that he had lost thousands of dollars by neglecting it—and there is probably no point in the whole range of grape growing upon which cultivators are so thoroughly agreed as this. The mere laying down the vines on the ground, covering them with snow, laying boards or brush upon or against them have all been found materially to increase the next year's product and to improve its quality. But these are clumsy expedients, incapable of systematic application and unfit for adoption on a large scale.

Where vines are trained to trellises in the manner which we have just described, it has been asserted by many that it is impossible to lay down the horizontal arms so as to cover them, owing to the rigidity of the old wood, and in order to avoid this it has been proposed to leave the head of the vine so low down that the arms shall lie on the surface and be always covered with earth. To this method there are many objections. The berries are soiled with every rain, clean culture is rendered more difficult, and the surface roots thrown out by the arms cause a succulent growth during moist weather, which suffers during the succeeding drought. But if the vines are bent down every year, little difficulty need be apprehended on this score, and if the following plan be adopted, vines may be bent sufficiently, even when they have become old and rigid.

The method which we have proposed, is to place the trellis 8 to 12 inches in advance of the vine, the stem being brought forward beneath the first slat or rail, and tied up as usual. The accompanying figure (17) explains this better than words can express it, and

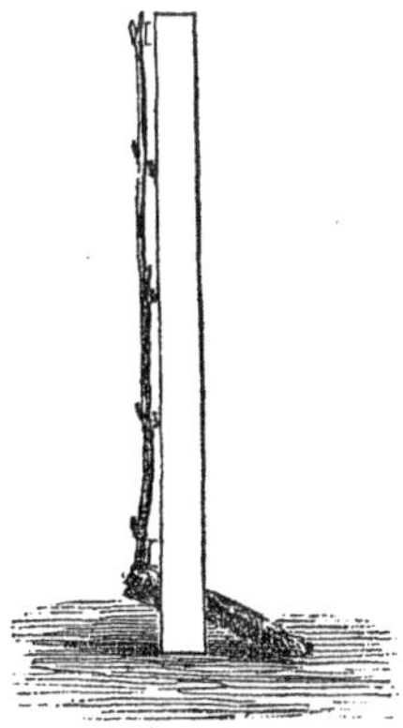

Fig. 17.

it will be readily seen that very little bending is required, and even that is so distributed over the

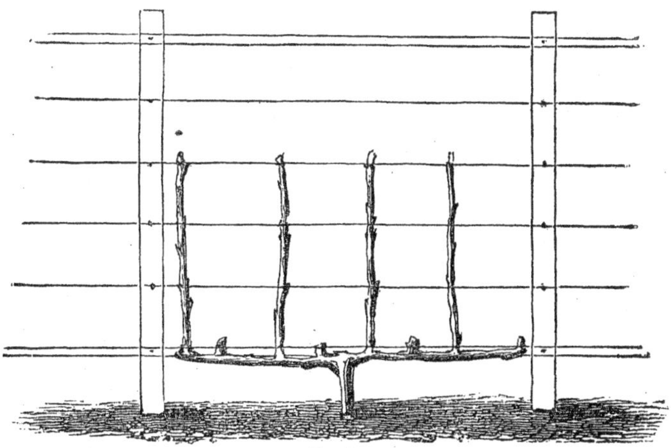

Fig. 18.

whole stem that no injury can result. No practical objections that we are aware of exist to this method.

Before bending down the stem, the vine should be pruned. This consists in cutting off the long shoots to a length of four feet (the first season), and the alternate short ones to the *lowest good* bud. The vine so pruned is shown in Fig. 18. Then the stem, having been bent down, it will be easy to fold the flexible young canes so as to lie compactly together, as shown in Fig. 19, when they may be

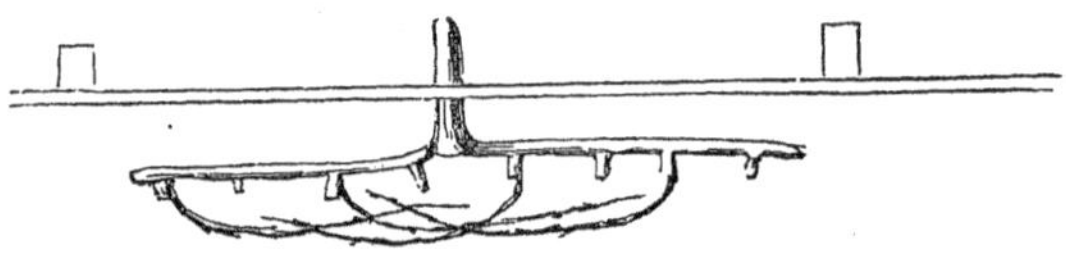

Fig. 19.

covered with earth. The soil for this purpose must be light and sandy, and should be so disposed that water will not penetrate to the vines. If light soil cannot be had, the vines may be pegged down and covered with the branches of evergreens, though it is improbable that these could be obtained in sufficient quantity to protect a large vineyard. Leaves or straw would answer, though they might harbor mice, which would soon destroy the vines.

The vines should be left covered as long as possible, but must be exposed before the buds begin to push in the spring. No particular day of the month can be given, the date varying with the locality and the season. The best mode of determining the point is to uncover some of the vines as soon as the cold weather has passed away. If they are swollen and ready to push, it is time to tie the vine to the trellis. If they seem still dormant, leave them a little longer. The later the vines can be made to push the better, as they not only escape late frosts, but their excitability seems to be so accumulated and intensified by such retardation that their after growth is much more vigorous than it would otherwise have been.

After the vines have been properly tied to the trellis, and the ground raked, or hoed level (all work on it being avoided when it is wet, however), nothing should be done until the buds have burst so as at least to show their vitality and strength. Then go over the vines and rub off all buds which show themselves on the upright stem and horizontal arms and disbud the canes so as to leave six good buds, and no more, on each. By doing this at this early period, the strength of the vine is thrown into the buds which remain, and they consequently push with increased vigor. The lowest good bud on the short spurs must also be left, all the others being removed.

As soon as the blossoms show themselves, and before they have expanded, it will be necessary again to go over the vines and stop or. pinch all the shoots which show fruit, at the same time removing all the blossoms except two or three clusters on each shoot. This will not only serve to keep the vine within bounds, but it will cause the fruit to set much better than it would do if this course were not pursued. In a former section, we alluded to stopping with a view to the ripening of the wood and the training of the vine, and the directions there given apply equally to our action as regards the shoots from the short spurs —they being designed to furnish the bearing canes for next year, to replace those which are now fruiting, and which will be entirely cut away at the next winter pruning. But other reasons also induce us to stop the fruit-bearing shoots, and as the whole subject of stopping, and its detrimental substitute, summer pruning, is one of vital importance to the grape vine, we cannot do better than preface our remarks by quoting the physiological laws upon which it is based, from Lindley's "Theory and Practice of Horticulture."

"Nature has given plants leaves, not merely to decorate *them* or to shade *us*, but as a part of a wondrous system of life quite as perfect as that of the animal kingdom. It would be of no use for a plant to suck food out of the earth by its roots, unless there was some place provided in which such food, consisting principally of water and mucilage, could be digested and so converted into the matter which maintains the health of the individual. The stem cannot do this: firstly, because it is a mere channel through which fluids pass; and, secondly, because many plants have no visible stem, as in the instance of the primrose; and yet in all such cases the plant feeds and must digest its food. It is to the leaves that this important office is assigned, and to enable them to execute it God has formed them with wisdom no less infinite than has been displayed in the creation of man. The leaves have veins through which their fluids pass and cells in which they are held while digesting, myriads of little caverns through whose sides respiration is maintained, a skin to guard them from the air, and pores for carrying off perspiration. A leaf is, in fact, both stomach and lungs; and to destroy it is to do the same injury to a plant as would be effected in an animal by the destruction of the parts to which those names are given. Of this we may be certain, that neither taste, perfume, color, size, nor any other property, can be given to a plant except through the assistance of the leaves; and that the more numerous these are, the larger and the more luxuriant, so, within certain limits, will be all that a plant is capable of forming. Strip the leaves off a tree, and no more wood will appear until the leaves are restored; feed its roots in the hope of thus compensating for the loss of its leaves, and the stem will be filled indeed with watery matter; but the latter will collect in the interior until it forces its way through the bark, and runs down in putrid streams, as happens to the mulberry-tree when it is incessantly stripped for silkworms, and as occurs to trees whose leaves are continually destroyed by a noxious atmosphere. Strip the ripening grapes of their green garments, and no color or sweetness will be collected in their berries. Rob the potato of its foliage and you will seek in vain for nourishment in its tubers; and so of all things else. On the other hand, leave the mulberry, the vine and the potato uninjured, to the genial influence of the sun and the air, and the dews of heaven, and wood is formed in the one case, sugar and color in the other—and flour, the staff of life, in the last, and these products will all be in exact proportion to the health and abundance of the foliage. . . . . .

"But although the general rule is to allow as many leaves to remain on a tree as can be kept in health, yet there are circumstances which justify their removal, and, indeed, render it necessary. For example, when a tender tree is trained to a wall, a great object with the gardener is to secure ripe wood; for unless he does this, the frost of the succeeding winter may destroy the branches, or the buds may be so imperfectly formed as to produce feeble shoots the ensuing season. To attain this object, those leaves must be removed which prevent the sun from striking upon the branches to be ripened, the effect of this being to stop the rapid growth of the branches and to consolidate their tissue, in consequence, partly, of the excessive perspiration, and partly of the rapid digestion of the sap, which is thus induced; *for the rate of digestion and perspiration in a healthy plant, is in proportion to the quantity of light and heat to which it is exposed.* Hence the removal of those shoots which in summer overshadow that wood of the peach-tree which is intended to be preserved another year, is useful; there can be no doubt, however, that as few shoots as possible should be thus removed. Another case in which the removal of the leaves is justifiable occurs in the vine. In this plant the fruit is borne near the base of the lateral shoots, which will, if unchecked, go on lengthening and producing leaves to a considerable distance. Now all the food of such a lateral shoot is obtained from the main branch, which, however, is only capable of furnishing a certain quantity. If the lateral shoot is allowed to grow unchecked, it will consume its portion of food in the production of many leaves and some grapes; and the more there is of the former, the less will be the weight of the latter. But if the shoot is stopped after having formed two leaves, all that quantity of food which would have been consumed in the production of other leaves is applied to the increase of size in the grapes, and the two leaves that are left; while on the other hand, the general crop of leaves on the vine will be amply sufficient to prepare those secretions which are to give flavor, color and sweetness to the grapes. This will, perhaps, be better explained by the annexed diagram.

"Let the line *a g* represent a lateral vine branch, bearing fruit at B, and leaves at *c d e f*. Suppose six ounces of sap are destined to support this lateral *a g*, during the summer; it is evident that, if equally distributed, each leaf and branch will receive one ounce of sap as its proportion. But if *e f g* are removed, it is obvious that the three which remain will have two ounces each, or double the supply.

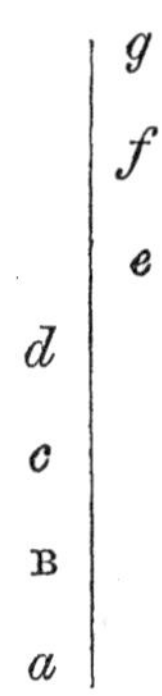

"Why, then, it may be asked, not remove *c* and *d* also? because, in that case, B, the bunch of fruit, would have the whole six ounces of sap to itself. The reason why this should not be done is this: if all the leaves on the lateral are removed, there will be no force left upon it wherewith to attract from the main branch the food that belongs to it; for the power which the parts of the plants possess of attracting fluid is in proportion to the amount of their perspiration. Now leaves perspire copiously, but the grapes themselves scarcely at all; whence their gradual conversion from a substance of the texture of a leaf into a mass of pulp. In the instance of vine pruning, the great object is to leave on the laterals just as much force as may be required to secure for the bunches the food that is intended for them, and at the same time to deprive the laterals of the means of expending that food uselessly in the production of leaves instead of fruit."

In applying the above to the culture of the grape in this country, however, we are inclined to believe that the direct access of the sun to the wood or fruit is not necessary to their perfect ripening. And our readers must also observe that, although in the illustration at the close of the paragraph, Dr. Lindley alludes to the "removal" of the leaves, yet from the remarks immediately preceding it, we gather that he is no advocate for "summer pruning," but for "stopping." By summer pruning we mean the removal of large quantities of leaves and shoots—a practice which is quite common throughout the country. Often and often have we seen loads of such matter cut away under the pretence of "letting in the sun and air to the grapes." Now if these summer pruners would only observe that all the finest bunches grow and ripen under the *shade* of the leaves, they would cease their senseless efforts and rest content with merely breaking off the ends of the shoots.

That grapes will not ripen well, and that vines will not be healthy under a dense mass of matted foliage, we freely admit. But this is not an evil to be remedied by the knife. In this case, most emphatically, prevention is better that cure.

When we reflect that the amount of organizable matter which can be furnished by any vine is limited, and also that all rank and succulent growth is prejudicial to the production of fruit, we can readily appreciate the advantage of directing the sap to the production of fruit, rather than wood and leaves. But we must also remember that every ounce of organizable matter which is embodied in leaves or stem, is so much capital invested, and is no more to be thrown away than the stock of the moneyed capitalist, which only brings in two per cent., even though his neighbor, on a different investment, receives ten.

The leaves are the laboratories in which the sap is prepared for the nourishment, not only of the fruit, but of the wood, and the more of them we have the better, provided we do not invest too large an amount of our available capital in their production, just as some of our farmers invest all their capital in land, and leave themselves nothing with which to work it.

Another evil attendant upon summer pruning, is the sudden and violent check which it gives to the plants. The roots being excited into vigorous action by the enormous draft made upon them, find themselves suddenly without a channel through which their unelaborated product can find vent; the balance of product and supply is upset and the fruit is filled with crude, ill-digested sap, thus causing it to be unripe and ill-flavored. But by early stopping the shoots, and thus preventing the further production of leaves and wood, we render summer pruning, that is, the removal of superabundant leaves and wood, unnecessary; no sudden check is given to the vines, the sap is fully elaborated as fast as it is supplied, and the fruit receiving an extra supply of properly prepared sap (which would otherwise have gone to the production of wood and leaves) is enlarged in size and improved in flavor.

That the leaves are the great agents in the elaboration of sap, was fully proved by the experiments of Hales, who forced orange flower-water into the vessels of a vine, with a view to impart its flavor to the fruit. The experiment was unsuccessful as to its ostensible object, but not as to its concomitant results; for he traced the flavor through the stem and branches to the leaves, but no further; there it was decomposed, and doubtless returned to the wood and fruit in the form of sap.

In a few weeks, or perhaps days, after being

stopped, the last bud on all these shoots, will, no doubt, burst and form a leader, which will grow nearly as vigorously as if the terminal bud had not been removed. It will, therefore, be necessary to go over all these vines again as soon as the fruit is set, and repeat the same operation. At this time, also, the fruit should be thinned, which, for vineyard culture, consists in the removal of all weak, ill-formed bunches, some even recommending the removal of the lower part of all the bunches.

When, however, extra fine bunches are desired, we prefer the plan usually adopted in hot-house culture, which consists in removing at least one half the berries from every bunch—the largest and finest being, of course, left. This operation is best performed when the grapes are the size of peas, but by many it will be deemed too minute and laborious an operation for vineyard practice.

While doing this, it will also be proper to remove or extirpate *all* shoots which either have not fruit, or are not wanted for next year's canes.

During the growing season it will be necessary to look over the vines, at intervals of two or three weeks, stopping the fruiting shoots, removing suckers, and pinching out laterals at the second eye. The ground should also be kept loose and mellow, and all the operations of the vineyard be carried on, with as little trampling on the borders as possible. Indeed, if the expense be not an objection, we would lay down boards or planks, supported by suitable stakes or posts, and forming a walk along the front of each trellis, so as to allow all the work of the vineyard being performed without a foot being set upon the soil.

CHAPTER VII.

SUBSEQUENT MANAGEMENT OF THE VINES.

The future management of the vines will consist in training up, each year, a shoot from the intermediate spurs, and cutting out entirely the cane which has borne the fruit. The cane which was trained up last year, will this year produce a crop, while, from the spur left in cutting out the former cane, is trained up a shoot for the following year, and so on *ad infinitum.*

As the peculiar pruning necessary is a subject of vital importance to success in grape culture, we will give a consecutive condensed description of it, illustrated by proper figures.

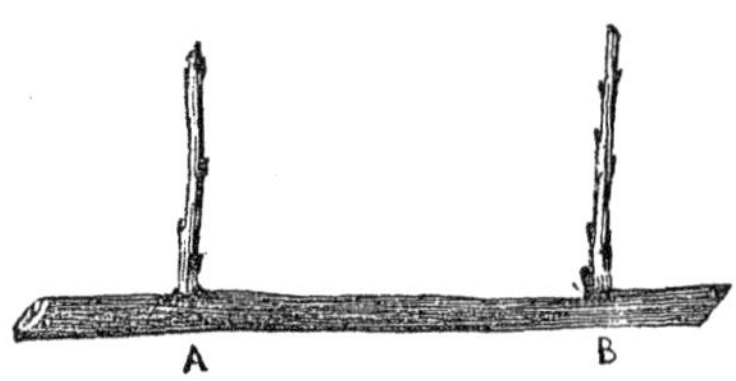

Fig. 20.

Fig. 20 shows a section of the horizontal arms, at the end of the third season. A is the cane which has been trained to the top of the trellis. B the shoot which was stopped when two to four feet long. Just before laying down the vine for winter protection, A is cut to about a length of 4 feet, and B is cut away at the cross-line, or just above the first good bud.

As the force with which the buds push, depends a good deal upon their number relatively to the size of the vine, it is absolutely necessary to cut off A to 4 feet or less, and rub out several of the buds which appear on it. If, in addition to this, all other buds except one from each of the spurs, B, be removed, we could scarcely fail to train up a good cane from B, even though none but latent buds were left.

Fig. 21.

Next season the figure is reversed. Here B is the young shoot of last year, while A, which carries the six shoots upon which the fruit grew, is cut off at the cross-line. B is shortened this year to 5 or 6 feet, and disbudded as before—one or two more buds being left on, as the vine is growing stronger.

At the base of A, below the cross-line, will appear intermediate little buds—some of them quite prominent. The best of them must be taken, and no fear need be entertained of getting a good cane from it, *if all the unnecessary buds are promptly extirpated.*

If, however, we allow shoots to grow all over the vine we will probably fail to get any cane at all.

The following season, the shoot proceeding from A

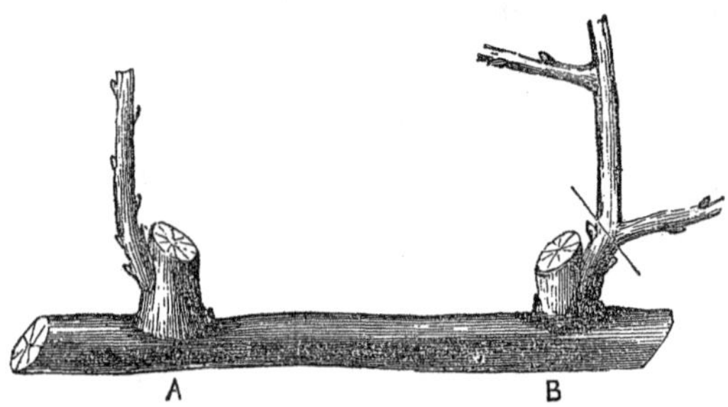

Fig. 22.

is fruited, and B is cut off at the cross-line. This stage of its progress is shown in Fig. 22.

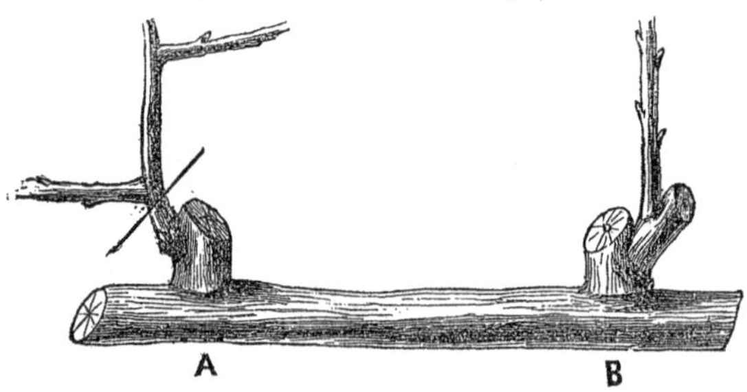

Fig. 23.

In Fig. 23 is shown the vine at the end of the sixth season. By this time, the spurs will have become hard, and if allowed to remain much longer, it will be necessary to renew the whole vine, as is done in the Thomery system (see Appendix). It will, therefore, be well to allow a bud to push from the base of B, if one should show itself, as there most likely

Fig. 24.

will. In this case, Fig. 24 will represent B as it will appear at the close of the season, when the entire spur must be cut off with a fine saw, at the cross line, and the wound carefully pared smooth and coated with a solution of shellac in alcohol.*

* "Take a quart of alcohol and dissolve it in as much gum shellac as will make a liquid of the consistence of paint. Apply this to the wound with a common painter's brush; always paring the wound smoothly first with a knife. The liquid becomes perfectly hard, adheres closely, excludes the air perfectly, and is affected by no changes of weather; while at the same time its thinness offers no

The shoot *a* is cut off at the line, as shown, or just above the lowest good bud. Next season, B will appear as shown in Fig. 20, and the same routine as that first described much be again gone over.

If we should be unable to obtain the shoot *a* at the time it is wanted (which, however, will not happen once in twenty times), we must leave the old spur and obtain a shoot from the base of last year's fruiting cane.

After a number of years (say six to ten), it may be found advisable to extend the vines. This may be done either by removing every second one, or by raising the trellis.

In the latter case it will be best, in order to secure an equal distribution of the sap, to lay down two courses of horizontal arms and allow the vertical, or bearing canes, to extend only half-way up the trellis.

The proper arrangement for this, is shown in Fig. 25, where it will be seen that the horizontal arms of every second vine are extended both ways, so as to cover double their usual space. The stem of the centre vine is carried up to the middle of the trellis and arms from it laid down, of the same length as the

resistance to the lip of the new bark that gradually closes over the wound. If the composition is kept in a well-corked bottle, sufficiently wide mouthed to admit the brush, it will always be ready for use and suited to the want of the moment."—*Downing*.

lower ones. The fruiting canes are produced and treated in the manner just described.

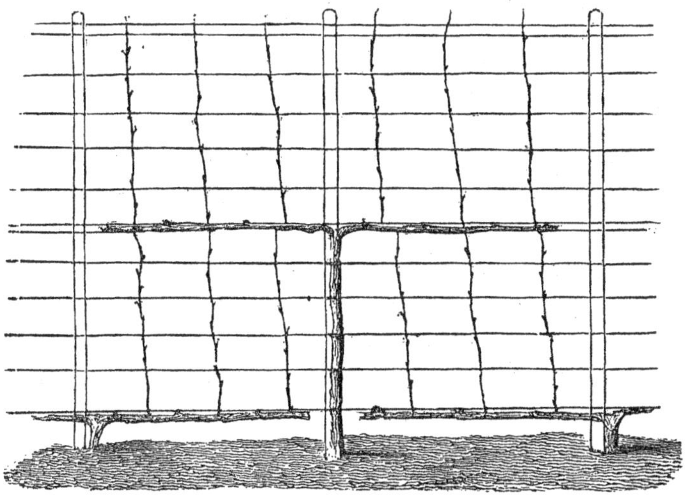

Fig. 25.

In order to effect the change, the lower arms may be extended by laying down the outer fruit canes of last year and pruning their junction with the old wood so as to leave a continuous rod. To produce the upper arms, however, it will, we think, be found best to cut the vines down to the ground and train up new stems, arms and verticals. The loss of time incurred will be more than repaid by the increased vigor and health of the vine.

That the general system of culture here laid down

is the best for all ordinary purposes, we are firmly convinced. The extent allowed to the vine during its first few years, is amply sufficient for the production of an abundant crop, while at the same time the vine is so far kept within bounds, that every bud is pushed with vigor. And this will be found to be one of the most important points connected with the proper training of the vine. For when the balance between the vital forces of the plant and the extent to which it is allowed to extend, is greatly disturbed, as exemplified in the opposite extremes of stake training and total neglect, nothing but debility on the one hand, and the inordinate production of wood to the exclusion of fruit on the other, can result.

CHAPTER VIII.

GENERAL THEORY AND PRACTICE OF PRUNING AS APPLIED TO THE GRAPE VINE.

In the preceding chapters we have given minute directions for that particular system of pruning and training, which we believe to be best adapted to our native grapes. It is now our purpose to detail those general principles which apply to *all* modes of pruning and training, and to describe a few of those peculiar systems which have been founded upon them.

I. The first principle upon which all correct pruning, whether of the vine or any other tree, must be based, is that the sap always tends to the extremities of the branches.

From this, it follows that unless the balance between the roots and branches of the vine be carefully and accurately adjusted, all the lower portion of the old wood will become devoid of spurs or bearing shoots, and unless the portion of the wall or trellis over which it is trained is otherwise occupied, the space will be left practically vacant.

Experience has also shown that there is no practical limit to this law—that is, that the distance to which the sap *may be* propelled exceeds any limits to which it is ever necessary to carry it.

"If the shoots of the vine are trained along a considerable extent of wall, the branches spread out much wider, and the berries attain a larger size. This property of the vine, although known to experienced gardeners, is not taken advantage of as it ought to be. A vine might be trained horizontally under the coping of a wall to a great distance, and by inverting the bearing shoots, the spaces between the other fruit-trees and the top of the wall could readily be filled up, and if different vines were inarched to the horizontal branch, the south wall of a large garden might be furnished with a variety of sorts from the stem and root of a single plant, the roots of which would not encumber the border in which the other fruit-trees are growing. I have an experiment of this kind now in progress in my garden. Within a few years past, I have gradually trained bearing branches of a small black cluster grape, to the distance of near fifty feet from the root, and I find the bunches every year grow larger, and ripen earlier as the shoots continue to advance.

"According to Mr. Knight's theory of the circulation of the sap, the ascending sap must necessarily become enriched by the nutritious particles it meets with in its progress through the vessels of the alburnum; the wood at the top of the tall trees, therefore, becomes short-jointed and full of blossom buds, and the fruit there situated attains its greatest perfection. Hence, we find pine and fir-trees loaded with the finest cones on the top boughs, the largest acorns grow on the terminal branches of the oak, and the finest mast on the high boughs of the beech and chestnut; so, likewise, apples, pears, cherries, etc., are always best flavored from the top of the tree. But I suppose there are certain limits, beyond which the sap would be so loaded with nutriment, that it could not freely circulate."

The sap being determined so powerfully to the extremities of the branches, the most unremitting attention is required upon the part of the vine-dresser, so that the bearing shoots may be equally distributed along the entire plant and an equal amount of nutriment directed to each.

But if through negligent management the bearing shoots or spurs are allowed to die out on the lower part of the vine, it will be difficult, if not impossible, to replace them.

By judicious pruning, the entire head of the vine may be so reduced that there will be abundant nutriment for all the buds, and by promptly and carefully stopping the more vigorous shoots, the sap may be so directed to the weaker ones that no difficulty of this kind need occur.

II. In this connection, we may consider a rule—which is laid down as a *principle*, however, by most arboricultural authorities—the buds are developed with greater vigor upon a branch which is cut short, than upon one which is left long.

This is true, but must be accepted with limitations. If there be two shoots springing from the same stem, one being pruned short and the other long, the buds on the long branch will be developed with the greatest vigor. If, however, the shoots be upon *different stems*, the buds upon that which is pruned most closely will push most strongly.

This we might anticipate, from the fact that there is more root power (if we may use the expression) to a given number of buds. But experiment would lead us to believe that if the lower buds are removed so that the same number of eyes are left on both, the longest would have the advantage—at least at first.

But as sap moves with greater difficulty through old than through young wood, the shortest shoot soon overtakes its companion and outstrips it. This principle is well exemplified in the rampant growth of those suckers which spring from old vines near the ground. They will frequently grow twenty or thirty feet in a season, while the strongest shoots at the extremities of the old branches do not exceed from five to eight.

Upon these principles is founded the rule which directs us to cut back plants which have made a weak growth, or have become old, gnarled and hard, so that they may throw up strong, vigorous shoots.

III. The sap supplied by the roots must be elaborated by the leaves, before it is fitted for the formation of wood or fruit, and the development of the roots is in direct proportion to the increase of leaves.

From this, it follows: 1st. That it is injurious to remove the leaves from the plant, with a view to ripen the fruit by the admission of sun and air (this point has been fully discussed, page 110, *et seq*).

2. That during the first two years' growth of the plants, the production of leaves should be encouraged as much as possible, so as to aid in the development of roots. Hence the plants should be carefully tied up, so as to preserve the leaves clean, active and uninjured, and abundant light and air should be furnished, so that they may be able to perform their part with efficiency. This being the case, it may be asked why we advise the stopping of the laterals which certainly form leaves, and hence must increase the growth of the roots. Two reasons may be given for the practice either of which are ample. The first is that it is not the leaves, *per se*, which do good, but their action on

the sap, to effect which, they must be supplied with air and light. Now, if the leaves on the laterals are allowed to grow, they crowd the foliage at the base of the plant, so that many of the leaves are partially shaded, while if these laterals are prevented from growing, the sap which would be absorbed by them goes to the elongation and enlargement of the main stem, by which the leaves are disposed over a greater surface and consequently maintain a more vigorous action. And, as during the first two years at least, the production of canes well ripened in their whole length, is no object, seeing that they are all to be cut away at the winter pruning, the vines should not be stopped, but should be allowed to grow to the end of the season. For as the roots require a certain degree of warmth to enable them to grow, and as the earth is in the best condition as regards temperature, just at the close of the growing season, it is best to maintain a vigorous action in the roots at that time—a time when they can make the most of it.

The second reason is, that the sap, as before stated, flows most vigorously through stout, free-growing shoots. Now, by removing the laterals, we increase the vigor of the main stem as well as its size, and hence not only obtain an enlarged, but a more suitable channel for the sap to flow in. The consequence is, that a well trained shoot will far exceed the aggregate of the same shoot and its laterals, if it be neglected during its growth.

IV. The more the sap is impeded in its course, the less vigorous will be the shoots produced, but the greater the tendency to bear fruit. This is exemplified in the pear-tree, where the branches are bent in order to produce fruit buds, and also in the common practice of bending the canes of the vine into bows and spirals, so that the buds may burst equally and produce fruit.

V. Whatever tends to diminish the vigor of the shoots and to force the sap into the fruit, enlarges the size and improves the flavor of the latter.

Upon this law depends the practice of summer pruning, which has been fully discussed in a previous chapter. And as it is necessary not only to diminish the vigor of the shoots, but to force the sap thereby saved into the fruit, the object of destroying all fruitless shoots (in bearing vines) is obvious, as well as the necessity of attending to the health of the roots.

Such are the general principles which should regulate the proper pruning of all trees; though they have been expressed chiefly with reference to the vine. In pruning with a view to the production of fruit, however, it is necessary to know the peculiar fruit-bearing habit of the plant under consideration. Thus upon the peach, fruit is always borne upon the last year's shoots; the pear bears its fruit upon spurs which have been formed during the previous year, upon old wood, and the fruit of the vine is always borne upon shoots of the current year, these shoots proceeding from either last year's shoots, or wood, which is much older. The last assertion is one which conflicts with the statements of most of our pomologists, and therefore it is incumbent upon us to give some evidence of its truth. Thus, Barry says:

"It must be observed, that the grape vine produces its fruit on shoots of the current year produced from eyes on the previous year's wood."

Du Breuil is more positive, and states that shoots which accidentally spring from old wood *never* produce grapes. His words are: "Dans la vigne, les grappes sont attachées sur des bourgeons naissant sur les sarments formés pendant l'été précédent. Les bourgeons développés accidentellement sur le vieux bois ne portent jamais des grappes."

We were rather surprised at this assertion, as it appeared to contradict our own observation. But lest it might be that the shoots which we had in view, had been produced by the remains of last year's rubbed out buds, we carefully watched a piece of old wood during one season, so as to assure ourselves that no buds had sprung from it between certain marked points. Next season the head of this old vine was pruned so severely as to cause several shoots to issue from the previously barren wood. Two of these bore fruit.

In performing this experiment, we kept carefully in view the difficulties attendant upon bringing it to a successful result, and although we succeeded in getting fruit from only two out of nine shoots, still, this was sufficient to establish the point. In performing it, care will be necessary to prune with sufficient severity to force the buds out of the barren wood, and yet to leave sufficient head to draw up the sap and prevent the too vigorous growth of the shoots after they are formed; otherwise the blossoms may change to tendrils. This experiment does not suggest any newer or better mode of pruning the vine, but it throws new light upon the laws which govern the formation of fruit buds, and exemplifies the fact that they are formed where the vital forces of the vine are so balanced that there is sufficient vigor and material to form fruit, and yet not so much rampant growth as to rob the blossoms of their necessary nutriment and convert them into tendrils.

That they are so convertible, every cultivator is aware, for it often happens that the hopes of the unskillful vine-dresser are disappointed—his fine show of blossom buds, turning out nothing but tendrils.

We believe the converse of this was first shown by Knight, from whose papers we make the following extract: "Every bunch of grapes commences its formation as a tendril, and it is always within the power of every cultivator to occasion it to remain a tendril. The blossoms are all additions, the formation of which is always dependent upon other agents; and if any considerable part of the leaves be taken off the branch prematurely, or if the vine be not subjected to the influence of the requisite degree of heat and light, the tendrils will permanently retain their primary form and office; and it is very frequently observable, when much of the foliage of fruit-trees has been destroyed, by insects, or when the previous season has been cold and wet, that blossoms are not formed at all, or are feeble and imperfect, and consequently abortive. . . . .

"The tendrils of the vine, in its internal organization, is apparently similar to the young succulent shoot and leaf stalk of the same plant, it is abundantly provided with vessels, or passages for the sap, and it is alike capable of feeding a succulent shoot or a leaf when grafted upon it. It appears, therefore, not improbable, that a considerable quantity of the moving fluid of the plant passes through its tendrils; and that there is a close connection between its vascular structure and its motions."

The various systems of vine pruning which have been founded upon these general principles, may be classified according to the part of the vine from which the fruit buds are produced. Thus, if we suppose A

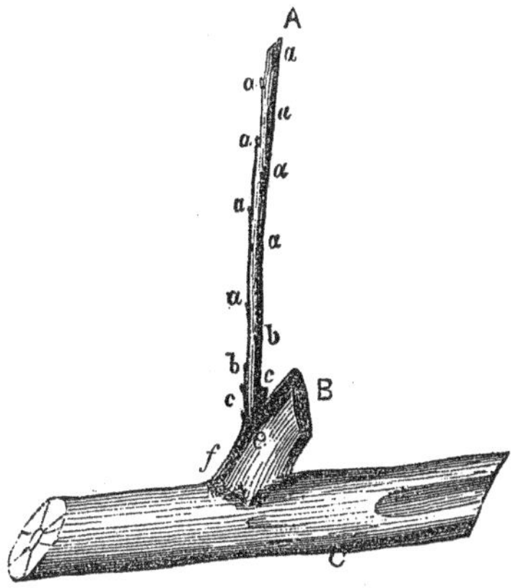

Fig. 26.

to be a shoot of last year; B a spur two years old, and C a branch three years old, then we may by judicious pruning obtain fruit, first from the plump buds *a a* on the young cane A; secondly from the buds *b c* near its base; thirdly from the buds *e* which will be found at its junction with the spur B; and fourthly from buds situated at *f*, that is at the junction of the spur B with the branch C. In the latter case, however, if the spur B is old and has borne several shoots like A it will require some skill and very favorable circumstances to procure fruit from the buds at *f*. But if B be always kept short, and no shoots be allowed except from its base, no difficulty of this kind will arise. Following this arrangement, then, we have:

I. The long-rod renewal system.

II. The long-spur system.

III. The short, or secondary-spur system.

IV. The close-cut, or primary spur system.

This system of classification we believe to be the only true one, although we are not aware that it has been adopted by any preceding author. Each of these four systems is not only distinct, but it may be adapted to almost any system of training, while none of them can be well combined in the same vine, unless the power of the roots is greatly in excess over the extent of the branches. Suppose, for example, on a vine with a well balanced head, a few long rods are left. No buds will start from the base of the spurs. But if all parts are treated alike, the eyes will break equally and in general will be all fruitful. The facility with which this principle may be explained and enforced is greatly increased by a clear and systematic classification, such as we have given above, and most authors have been aware of the importance of such a classification; but if the reader will compare the attempts of Loudon, who depends in his classification, not only upon the system of pruning, but of training; of McIntosh, whose three systems are "the spur," "the long-rod," and "the irregular" forms; of McPhail, who has the "fruit-tree method;" "the spurring-in system" and "the long-rod sytem;" of Von Babo, who has "head pruning," "limb pruning," "frame pruning," and several sub-varieties named after the localities in which they have been adopted; and most of our American authors, who have simply the renewal and spur methods, with that given above, we think he cannot but give his preference to the latter. And as all systems of pruning with which we have ever met may be easily referred to one or the other of our four classes, we will describe them in detail and give a few illustrations of each, as derived from the practice of our best growers.

I. The long-rod or renewal system is generally attributed to Clement Hoare, who adopted it in his "Practical treatise on the Cultivation of the Vine," and as he has not seen fit to give the credit of it to prior authors, most of his readers have awarded it to him. But it is substantially the "new method" of Switzer; the alternate system of Speechly, and the "new and experimentally proved superior method" of Kecht. It is certainly very old, though it is still commonly called the "new method."

The system which we have adopted as the best for

vineyards and gardens in the northern States *where our native vines are cultivated* is substantially the system explained by Hoare. The system pursued in Ohio and in many European vineyards, is also a modification of the long-rod system, but as we propose to give a full accouut of Ohio vine culture amongst our examples of American vineyard practice, we need not dwell upon it here.

The following are a few of the most elegible modifications which have been proposed:

Mr. John Mearns, in the Horticultural Transactions, (vol. iv.) describes a system which is not only well adapted to the hot-house culture of the vine, but is one of the best with which we are acquainted where it is desired to fruit quickly, a great variety of grape vines in a small place. This method is as follows:

"My method of managing vines is in some respects different from any other with which I am acquainted; by it I have never failed, for the last eleven years, to obtain invariably the same luxuriant crops, although I have never allowed above one-third of the bunches which showed themselves to remain on the vine; and each succeeding crop has been as uniform as if the branches had been placed, artificially, over the whole roof. I have no doubt but, under the same treatment, the vines will continue to be equally productive for any length of time. The shoots are so vigorous that their girth is, generally, at the end of the season, from an inch and a half to an inch and three quarters. The branches, in their most luxuriant growth, never appear in any confusion, even to those who are but little skilled in the cultivation of grapes, and the method is so simple, that it may be described with the assistance of figures, so as to be perfectly comprehended by any person in the least acquainted with the nature of the vine. I have never deviated from it since I planted the vines in the spring of 1806.

"My vines were planted two feet and a half apart, and being watered to settle the earth round their balls, I headed them down to within a foot of the soil, as here represented.

Fig. 27.

"I only allowed one shoot to proceed from each plant the first year; rubbing off all the others before they had completely burst into leaf, the uppermost being the one I retained. In the course of the summer I watered them with soft pond water, as I found they wanted it, and frequently with drainings from the farm-yard, and with soap suds, when I could procure any.

"During the first summer, the vines made quite as

much progress as I could have expected, and their different degrees of vigor were nearly in proportion to the state of the roots when planted. When the leaves had fallen in the end of the year, I cut them down to the second or third eye, when they had this appearance. (Fig. 28.)

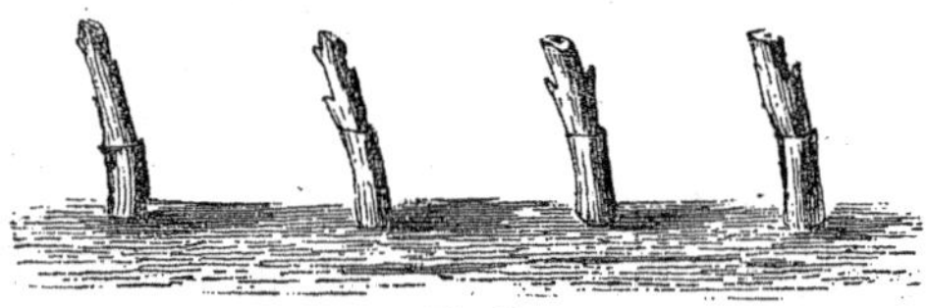

Fig. 28.

"In the beginning of the succeeding February, I excited them gradually into action by a little fire heat, and when the buds were ready to burst I rubbed all off but the two finest on each plant; the strongest of these I intended to furnish bearing wood for the lower half of the roof for the following year. The most feeble of the two was cut down to the second or third eye, at the end of the season, and at the same time the strongest shoot was reduced to eight feet, being the length of the lower half of the rafter.

Whilst they were growing during the second summer, I kept the shoots regularly trained upward, divesting them of tendrils and laterals. I only allowed the strongest of the two leading shoots to run about three, four or five joints beyond the middle of the roof (where I intended to cut them at their winter pruning), according to the vigor of the different shoots; and then I pinched off their tops, in order to strengthen the eyes for the ensuing season. The weaker shoots I only suffered to run about three four, or five feet, according to their strength, and I then pinched off their tops, never allowing them to push above two or three eyes from the same place, during the remainder of the season, without pinching them back; and then retaining only a single eye, unless I found it necessary, in consequence of the vigor of the

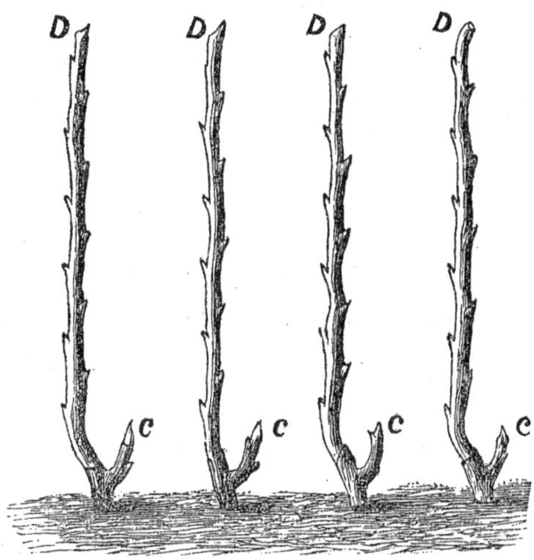

Fig. 29.

vine. I kept the laterals stopped back also to the first leaf. At the fall of the leaf, I cut the leading shoots at the middle of the rafter, and the lower one at the the eye, as is here represented. (Fig. 28.)

"The preceding sketch represents four separate vine-plants, at the end of the second season after they had been planted, when the strongest shoot had been headed down to the middle of the rafter D, and the weakest shoot to C.

In the third season, I carefully preserved the uppermost shoot from the end of my bearing branch at D, as a leader to furnish the upper part of the rafter with bearing wood for the next year; and I also trained upward the leading shoot from the bottom spur C, which I intended should become the bearing branch for the lower half of the roof in the following season. I was careful that none of the tops of these leaders should meet with accident, till they had reached their destination for the season—that was about three or four joints beyond where they were intended to be cut down, to the winter pruning. All the buds on the bottom spur C, were rubbed off, except the leading one. As I bore in mind the necessity of a bottom spur to produce a succession shoot from the bottom in the following year, which was necessary to the regularity of the system I contemplated, I selected one of the most convenient buds for my purpose, from the bottom of the old stems, all of which were now putting out several buds; but I suffered none except the selected one, to remain long after it had made its appearance. The management of the young shoots of the year was, in this and the following seasons, the same as I have before detailed.

"In the autumn of this, the third season, the lower half of the house was furnished with a crop of ripe grapes upon the wood of the preceding year, and parallel to it on each vine grew a young shoot, intended to bear the lower crop the next year; whilst the upper half of the house had single shoots trained from the end of the bearing wood, which shoots were also to bear a crop the next year; and besides these, a third shoot on each vine had been trained from the bottom bud, which I had not removed, and which were about four feet in length, having been treated as the weaker shoots in the second year's management, which I have described, and to which they were similar. When this half crop was gathered and the leaves had dropped, I cut off the top leaders level with the uppermost wire of the house to which they were tied, and the lower leaders level with the middle of the roof (the top and bottom leaders, or bearing wood for the next season, being each eight feet long), and the bottom or weak shoot, above described, was cut down to the second or third eye, as the lower shoot had been cut in the preceding winter. All the spurs of the lower

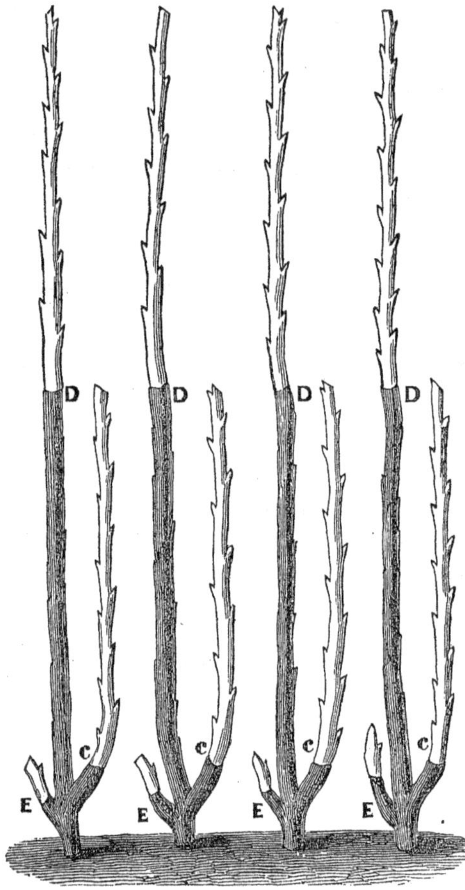

Fig. 30.

part of the shoot, which had now reached the top of the house and had borne the crop of grapes, were cut clean out. The following was the appearance of the same four vines, after they had been pruned in the third winter, when they were in a state to produce their full crop in the following season. (Fig. 30.)

In the fourth summer a full crop was produced both in the upper and lower half of the house; the longer shoot D bore its bunches on the upper half of its length, and it was not suffered to extend itself by a leading shoot; the shorter shoot C bore its bunches on its whole length, and extended itself by a leading shoot to the top of the house; the spur E was suffered to become a shoot, extending a few joints beyond half the length of the rafter, and from the bottom of the old wood a weaker shoot, as before, was trained, to become the foundation of the lower shoot of the next season. In the pruning season, D, which had become the longest branch in the previous winter, was entirely cut away from the bottom; the shorter branch C, which had now become the longest, was stripped of its spurs on its lower half of the old wood, and its upper half was left for bearing; the extended spur E, became the lower bearing branch, and the weak shoot F (Fig. 31.) at the bottom, was reduced to a spur, to furnish the lower wood for the next year.

The following figure represents the plants after

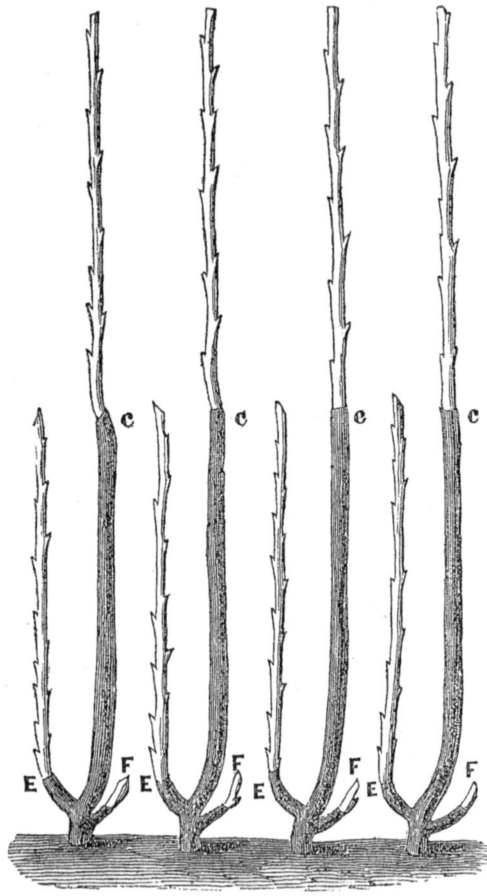

Fig. 81.

being pruned the fourth season, the sides being reversed.

With this alternation of pruning, the system has been continued to the present time, and may continue as long as it shall be desirable to have the house in bearing.

During the last four years, I have stopped the bearing branches at the bunch, instead of the next joint above it, which is the usual practice; for I found that the fruit did equally well and it divested the branch of an incumbrance, while it allowed a much larger portion of light to come into the house, together with a more free circulation of air among the fruit and young wood.

I blind all the eyes on each fruit spur as soon as they push, except the uppermost, which I retain to draw up the sap to nourish the fruit. I never suffer them to push above a joint or two, before I pinch them back, always cautiously retaining an eye. By constant stopping, the eyes soon increase to a large cluster, when I frequently find it expedient to pinch out a great part of them with my finger nails, unless I see danger of its exciting my next year's fruiting eyes to burst prematurely. I am particularly cautious that nothing shall happen to injure the leaf that accompanies the bunch, for if that is lost, the fruit, of course, will come to nothing.

During the summer I inspect the vines regularly every morning; seeing that the ends of my leaders

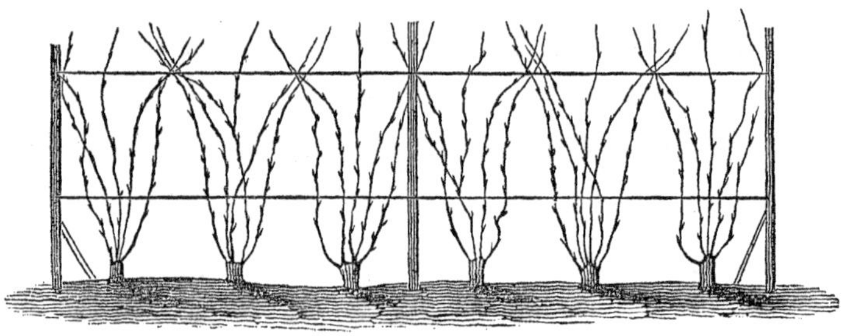

Fig. 32.

are in their proper places, and not obstructed; picking off tendrils and stopping the laterals above the

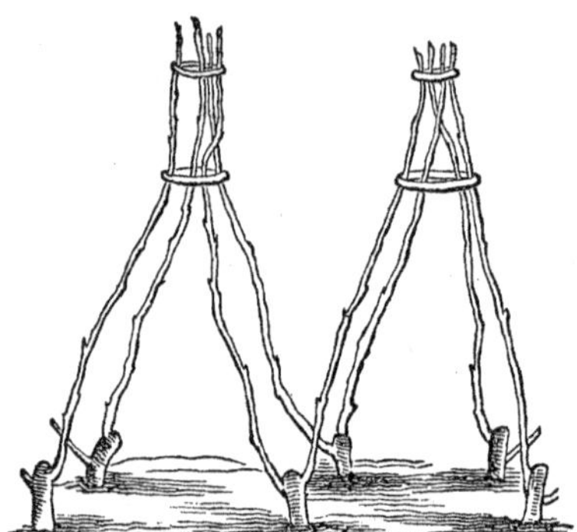

Fig. 33.

first leaf, on my next year's bearing wood, tying down fruit spurs carefully, and stopping any shoot that may have sprung from the ends of them; as well as other shoots that may come out from the previously stopped laterals.

Fig. 32 shows a simple method of training wines to a trellis formed of light stakes or a couple of

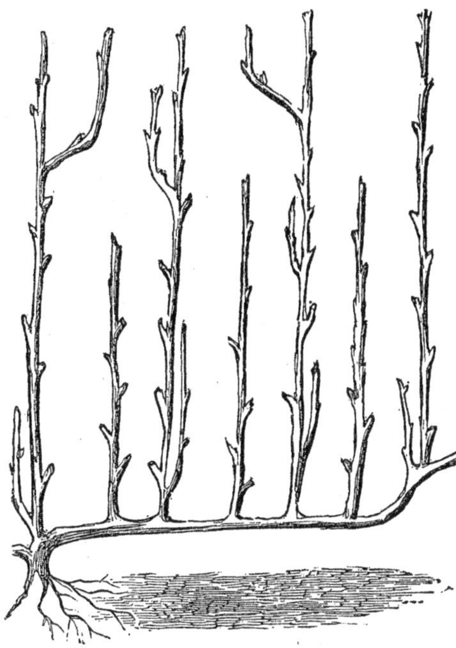

Fig. 34.

wires. If trained on the plan shown in Fig. 33 neither stakes nor trellises will be required.

II. The long-spur system is that upon which old vines are trained, and consists in cutting the young wood the previous year, back to three or four eyes, all weak shoots and dead wood being removed.

Sir J. Paxton, in the "Gardener's Chronicle" for 1842, gives the following directions for pruning vines on the spur system. The cut there given (Fig. 34) has been often reproduced, but in general it has been so reduced that the character of the shoots is not clearly seen:

"It represents a portion of the vine when pruned in autumn, on the spur system, with short rods of five or six eyes each, left at convenient intervals on the oldest branches throughout the vine. The perpendicular main shoots should not be less than two feet apart, and when pruning them no useless eyes should be left, that is, no eye should be allowed to remain but where a shoot is desired in the following season. By attending to this, the vine will not have to develop (as is usually the case), an immense quantity of superfluous branches; and although the operation may appear a tedious one at the time of pruning, an immense saving of labor and time may be effected at a busier period in the spring, and the quantity of fruit may be easier regulated in proportion to the strength of the vine. If this is attended to, nothing will be required in the summer but securing the young fruit-bearing shoots to the wall, and shortening them to one joint above the bunch as soon as the fruit is set, excepting the leading shoots, which should not be stopped until the lower part is ripened; otherwise the main eyes for the next season may be induced to grow prematurely. In autumn the young wood from the spurs is shortened back to one, or at most, to two eyes, and the terminal shoots in proportion to their strength; but for the strongest wood, from eight to twelve eyes will be found as many as will break well.

"When commencing to train a young vine in this manner, the side branches should not be brought to the horizontal position at first, but lowered gradually as the number of suitable branches for upright stems are obtained; by this means they acquire strength faster than if trained horizontally at first."

It is obvious that this system is nearly the same as the long rod, or renewal system—the difference being that instead of taking several of the upper buds on each young cane, we use only one and have a great many canes or spurs.

The only real advantage to be derived from it (so far as we are able to judge), and that upon which its distinctive features is founded, is that the buds from which the next year's crop is to be obtained are always well ripened. We would, therefore, prefer it to the first system, where the vines are tender, or the climate unfavorable, and deem it of sufficient importance to give in detail a method of treating the spurs during a series of years.

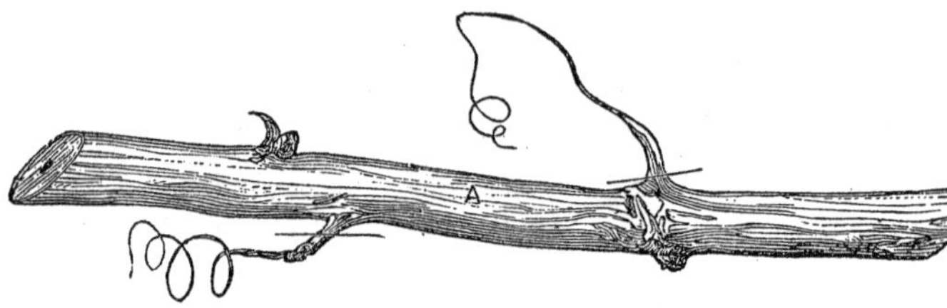

Fig. 35.

Fig. 35 (A) shows a portion of a young cane which may either form part of the vertical branches on a trellis, or the single stem of a young vine. The first season of fruiting, the tendrils should be cut off and the buds thinned to from six to ten inches apart, depending upon the vigor of the variety; and so that they will be alternately on different sides of the cane, thus leaving the buds on each side from 12 to 20

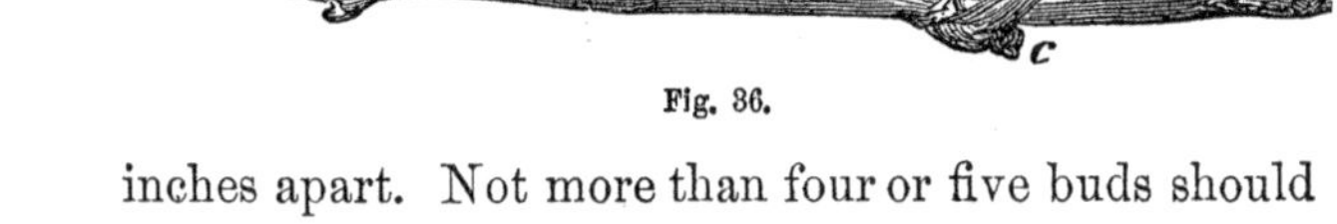

Fig. 36.

inches apart. Not more than four or five buds should be left on a cane during the first season.

At the close of the first season, after the leaves have fallen, the cane will present the appearance shown in Fig. 36. Here A is the main cane; B is the shoot produced by the buds on Fig. 35 ; and C is a bud at the base of this shoot. Prune the shoot B to one plump bud, as shown in the figure and allow the bud C to push and form a shoot; stopping it, however, as soon as it has made a few leaves.

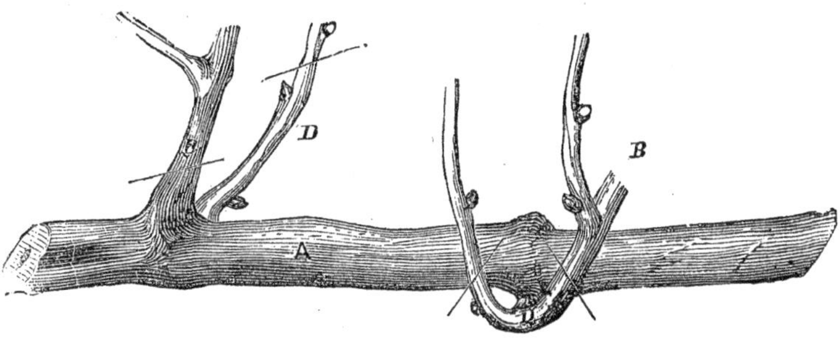

Fig. 37.

Next season we will have the shoot D, (Fig. 37) with several nice, plump buds, and the old spur B, with its shoot which bore fruit last year. Cut D back to one or two eyes, and cut B away entirely. The buds on D will push and bear fruit, and a bud will, no doubt, push from the base to form the spur for next year.

Fig. 38 shows the next winter pruning. From this description, it is obvious that we must, each year, have eyes to produce, not only fruit, but a young cane, which will form the spur for next year. If we depend for this spur upon last year's fruiting

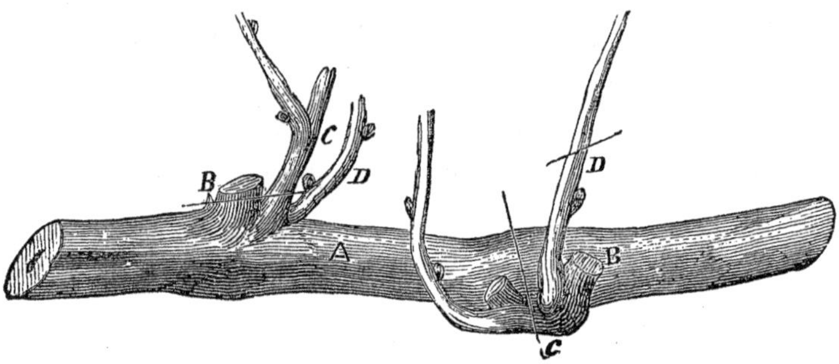

Fig. 38.

shoots, our spur will soon become so long, and our vine so encumbered with old wood as to be quite unmanageable, unless we adopt the system to be next described:

III. Here we depend for our fruit upon buds proceeding from the base of last year's fruiting shoot, this fruiting shoot being borne upon a spur attached to the main branch. This is the system of pruning adopted at Thomery, and as no good description of it is be found in any American publication with which we are acquainted, we give the very full and lucid account by M. Dubreuil—a translation of which may be found at the close of the volume.

IV. In the short-spur, or Thomery system, the fruit-bearing shoot proceeds from a spur on the main branch, which although short, is still a spur. Theory, however, would lead us to suppose that it might just as well proceed from the juuction of last year's fruit-bearing shoot with the main branch.

CHAPTER IX.

WALLS AND TRELLISES—THEIR INFLUENCE AND CONSTRUCTION.

Although the influence of the various forms of walls, trellises and stakes upon the growth and maturity of the vine depends somewhat upon the system of pruning and training pursued in connection with them, still, it cannot be doubted but that their forms and the materials of which they are made also exert an influence which is by no means to be disregarded.

In this country, walls devoted to the culture of the vine have not been used to a sufficient extent, to afford reliable data as to the benefit to be derived from them. Many single vines, however, are trained on the ends of houses and along board fences, and from a careful examination of several such examples, we are inclined to believe that in exposed situations the erection of cheap walls would pay well, even in vineyards.

When vines are judiciously trained in front of brick walls and *at a few inches' distance* from them, the grapes uniformly ripen sooner than those on exposed trellises. The wood also is more perfectly matured, and this, during a succession of years, exerts a considerable and favorable influence on the vine.

The effect of walls doubtless depends upon two causes, one being the higher temperature produced by the radiation from the surface of the solid wall, and the other being the protection from wind and storms which such a structure affords.

That the latter point is one of material importance, we are well satisfied, for however essential ventilation may be to the healthy growth of the vine, all violent winds and cold blasts are to be studiously avoided.

A striking instance of this is to be seen in the garden of a gentleman of this city. Several vines are there trained along the east side of a high board fence, and although the same judicious and systematic care is given to all parts of the vines, yet the finest fruit is uniformly found a foot or two below the top of the fence. Now when we remember that on all open trellises the finest grapes are found at the *top*, since all trees produce the best fruit at the *extremities* of the branches, we must attribute no mean effect to the protection afforded by the fence, since the boards of which it is composed can scarcely be supposed to retain and radiate much heat, and its height (about eight feet) is not sufficient to include the limit to which vines may be judiciously carried.

"The actual temperature to which a tree trained upon a wall facing the sun is exposed is much higher than that of the surrounding air, not only because it receives a larger amount of the direct solar rays, but because of the heat received by the surrounding earth, reflected from it and absorbed by the wall itself. Under such circumstances the secretions of the plant are more fully elaborated than in a more shady and colder situation, and by aid of the greater heat and dryness in front of a south wall, the period of maturity is much advanced. In this way we succeed in procuring a Mediterranean or Persian summer in these northern latitudes.

"When the excellence of fruit depends upon its sweetness, the quality is exceedingly improved by such an exposure to the sun; for it is found that the quantity of sugar elaborated in a fruit is obtained by an alteration of the gummy, mucilaginous, and gelatinous matters previously formed in it, and the quantity of those matters will be in proportion to the amount of light to which the tree, if healthy, has been exposed. Hence the greater sweetness of plums, pears, etc., raised on walls from those grown on standards. It has been already stated that an increase of heat has been sought for on walls by blackening them, and we are assured in the 'Horticultural Transactions' (III. 330) that, in the cultivation of the grape, this has been attended with the best effects. But, unless when trees are young, the wall ought to be covered with foliage during the summer, and the blackened surface would scarcely act, and in the spring the expansion of the flowers would be hastened by it, which is no advantage in cold, late springs, because of the greater liability of early flowers to perish from cold. That a blackened surface does produce a beneficial effect upon trees trained over it is, however, probable, although not by hastening the maturation of the fruit; it is by raising the temperature of the wall in autumn, when the leaves are falling, and the darkened surface becomes uncovered, that the advantages are perceived by a better completion of the process of growth, the result of which is the ripening the wood. This is indeed the view taken of it by Mr. Harrison, who found the practice necessary, in order to obtain crops of pears in late seasons at Wortley, in Yorkshire (see 'Hort. Trans.' III. 330 and VI. 453.) It hardly need be added that the effect of blackening will be in proportion to the thinness of the training and vice versa." —*Lindley.*

The articles referred to by Lindley, being short and practical, it may be well to transcribe them. Henry Dawes writes thus to Sir Joseph Banks: "I take the liberty of communicating to you my remarks on a garden wall, on which I have been making experiments at Slough. It faces the south, and against it, about the middle, a young grape vine is trained. Two years ago I covered a portion of the wall with thick black paint. The vine was divided into two equal parts, one half was trained on the painted, and the other on the plain wall. The season was so unfavorable last year, that scarcely any out-door grapes came to perfection; but those in the blackened part of the wall were much finer than those on the plain part. This year the success of my experiment has been complete. The weight of fine grapes gathered from the blackened part of the wall was 20 lbs. 10 oz., while the plain part yielded only 7 lbs. 1 oz., being little more than one-third of the other. The fruit on the blackened part of the wall was also much finer, the bunches were larger and ripened better than on the other half; the wood of the vine was likewise stronger and more covered with leaves on the blackened part.

"It is a generally known fact, that a black, unpolished surface absorbs more rapidly than other colors the sun's rays, and thereby becomes sooner heated. It is equally well known that surfaces which absorb heat more quickly, part with it more easily when the source of heat is withdrawn, and cool quicker. In the summer time, when the days are long, the wall will be more intensely heated under the blackened surface, and the night (or time of cooling) being short, it may not have returned to the temperature of the air, before it is again subjected to an increase of heat. If the time of cooling were long enough, that part of the wall under the blackened surface, might become actually cooler than the part not blackened, and thus the extremes of heat and cold be greater than when the wall was left with its usual surface. In the summer time, however, the wall is not only more intensely heated, but probably retains a great portion of the heat during the night.

"Horticulturists will decide which of these two causes is efficient in producing the effect I have stated, or whether both may not coöperate; it is not for me to presume to do so, though I should be inclined to think, that in this climate, the intensity had more influence than the uniformity."

Chas. Harrison, gardener at Wortley Hall, Yorkshire, gives the following directions for blackening walls:

"When the leaves have fallen in the autumn, I take the earliest opportunity to loosen the tree from the wall and to prune them; the wall is then colored with coal-tar, mixing with every gallon of the tar one pint of linseed oil, in order to prevent it having a shining surface when dry. It is more necessary to make this addition in the hotter parts of the kingdom than it is here, but even here it is essential in hot summers, for when the sun shines strongly on the wall with a shining black surface it has appeared to me to scorch those shoots which touch the wall; but this does not happen when the color is rendered opaque by the mixture of the oil as recommended. If the wall has not been previously colored, I give it a second coat as soon as the first is dry. In laying on the color care is taken that the liquid is not sprinkled upon the trees, for it would close up the pores of the wood and consequently do injury.

"After the wall is colored I allow the trees to remain loose from the wall until the coal tar has set (unless strong winds prevail, in which case I secure the main limbs and branches to the wall), in order that the shoots may not be damaged by coming in contact with it before it is dry. When the wall has become moderately dry, I nail the trees to it. A wall of sound bricks will not require recoloring more than once in ten years. Coal tar being very cheap, a wall of considerable extent may be colored for a trifling sum. Any dark-colored paint will answer the same purpose, but it is far more expensive, and requires renewal more frequently.

"The dark color, absorbing the rays of the sun, the wall acquires at least ten degrees of heat more than the walls not colored, as directed; thus affording great assistance in maturing the buds upon fruit-bearing shoots, so that the trees may be productive. In cold and wet seasons, without such aid, I should not have been able to obtain ripe buds upon fruit-trees under my care. This I have had ample proof of by the unfruitfulness of those trees which are against walls not colored, at the same time that trees against colored walls were abundantly fruitful. The wall being colored is also a preventive of insects harboring in it and also tends to keep it dry.

"The growth of young trees is much promoted by the coloring and they are sooner brought to a supply of fruitful buds."

In all cases in which vines are trained in front of walls or fences, it is important that a space of from six to twelve inches be left between the wall and the trellis to which they are fastened. If trained directly to the wall, the vine will not only be subject to mildew, but ventilation will be materially interrupted. The bunches also, are liable to injury when lying against the surface of the wall.

Walls may, of course, be constructed of any material, brick, stone or concrete. Brick is probably the most suitable material, though, as plain walls can be rapidly and cheaply built of concrete, it is probable that it might pay to erect them on an extensive scale in some parts of the country. In the celebrated Thomery vineyards, the walls are built of clay with a cap of thatch. It is probable that walls built of well made sun-burnt bricks would last a long time and answer a good purpose if properly protected by a cap or eave of board or straw.

But, for all practical purposes, our reliance for vineyard training, in the present state of our experience, must be upon properly arranged trellises. We will, therefore, give what we consider the best mode of constructing them.

If the vines have been planted two years previously at distances of eight feet in the rows and the rows six feet apart, the first step to the erection of the trellises is to set up a post between each vine and slightly in advance of the rows, so as to facilitate bending the vines for winter protection.* These posts may be of such size and material as the vine dresser may procure. Cedar, chestnut, locust or oak make the best, and a good size is four inches deep (across the rows) and three inches thick. They should stand from seven to nine feet out of the ground and be sunk not less than two and a half feet—if three feet, all the better. The two posts at the ends of the rows must be placed so that they cannot be drawn inward. Various devices for effecting this are shown in Figs. 39 and 40.

* See page 107.

One consisting simply of a piece of plank nailed across the post so as to afford a broad surface to lie against the earth. The other is secured by a brace,

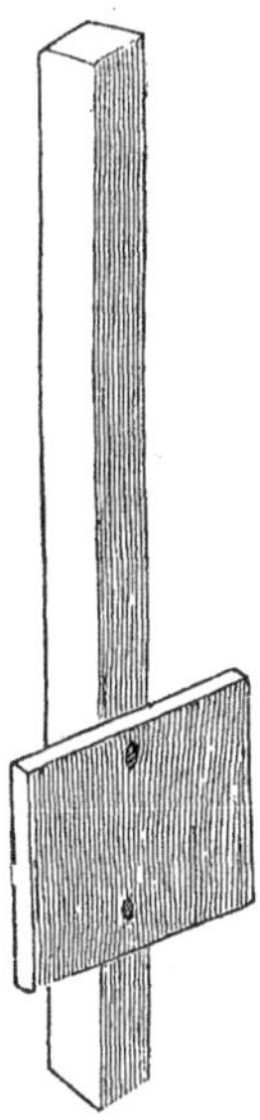

Fig. 39.

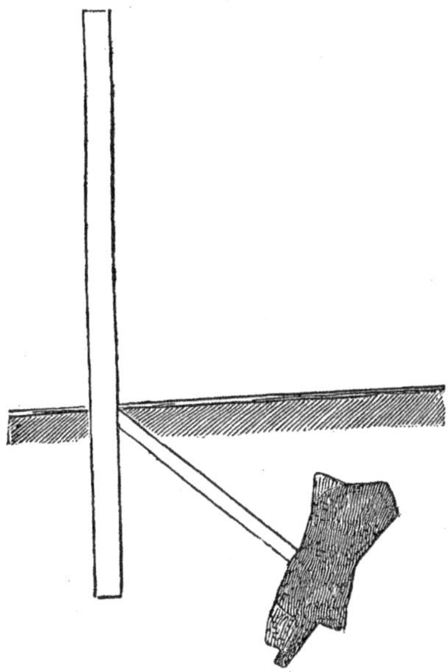

Fig. 40.

which rests against a large stone sunk below the surface.

The posts having been set in a straight line and reduced to a proper height, the next step will be to nail two strips of wood, one along the top and the other at from 9 to 14 inches from the ground, or just at such a height that the head of the vine-stems (from which the horizontal arms spring) may reach its upper edge when laid against it. If the trellis is over seven and a half feet high, it will be well to nail a third slat, equidistant between the two, though it is not absolutely necessary. The next step is to divide the spaces between the slats into equal parts of about 15 inches each. Thus if the trellis be seven and a half feet high, and two slats (the lower one ten inches from the ground) we would divide into five spaces of 16 inches each. If nine feet high with three slats (the lower one twelve inches from the ground) we would divide each of the two spaces into three divisions of 16 inches each. Then take No. 12–16 annealed iron wire, twist a good loop on the end, and having slipped it over a stout nail driven into the end post, draw the wire along the posts, attaching it to each with a small staple well driven in until the last is reached, when the wire may be twisted round a nail or pin and the loose end secured by a staple.

Various devices have been proposed for drawing the wire tight and adjusting it for contraction and elongation according to the temperature, as is done in the construction of fences. But we are satisfied that there is no necessity for this, as the wire can be drawn over a stretch of eight feet tight enough for all practical purposes, while it will always be loose enough to allow of any contraction that can take place. The truth is, that the wires do not require to be so very tight; even if they do have a little motion from the wind, it is not productive of any injury.

Our method of putting up the wire is as follows: We first provide a pair of strong wooden pincers such as those shown in Fig. 41, the handles of which are at least 30 inches long, and having a piece of stout sole leather tacked over the jaws.

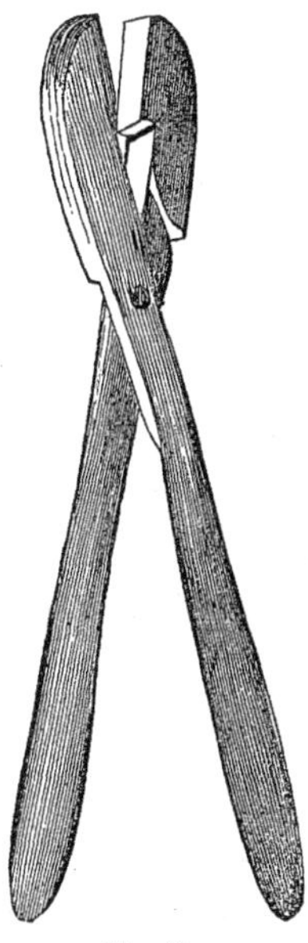

Fig. 41.

Then having secured one end of the vine to the first post and uncoiled the roll, laying it on the ground, in front of the posts to which it is to be fastened we grasp it between the leather jaws of the pincers and step slowly back, straining it as much as possible until we are past the second post, when an assistant fastens it firmly with a small staple and we are again ready to step back to the next. By means of this contrivance, the wire can be laid on as tightly and smoothly as possible, for all the kinks are taken out by passing it through the leather jaws of the pincers, which should be well greased. It is necessary to go into the field provided with several sets of leathers, as they soon wear out, but are easily renewed. They should be at least three inches broad, so as to straighten out the wire thoroughly.

That wire is better than wooden slats there can be no doubt. It is less in the way, the vines cling to it more readily and the appearance is vastly superior. A pound of No. 12 wire will stretch across three posts (24 ft.) so that a trellis 8 feet high with two wooden slats and five wires, each 14 inches apart, will require 1⅔lbs., which, at eight cents per lb., will cost about 13 or 14 cents. As the cost of the wire is considerable, some cultivators do with less. Dr. Underhill's trellises are seven feet high with only three wires, and we believe no slats. But we prefer the arrangement just described, as we can thus tie in each shoot conveniently and regularly, and the wires at 14 inches are none too close to have a bearing shoot on each. When wire cannot be conveniently had, the following is a good mode of putting up a rough trellis.

Set the posts as usual, and provide a number of slender split poles (hoop poles) and also a sufficient number of wire staples made of strong wire (No. 6 or 8 hard). Then secure the poles or slats to the posts by means of the staples, the ends of the poles being made to lap over each other so that two may be fastened by one staple. The accompanying figure (42) will explain this better than words can describe.

Fig. 42.

If the staples are a *little* less than the poles, the latter will be held very firmly. Nails will not answer, as they are apt to split the poles, and we are inclined to think that, for ordinary slats, staples would be better than nails on this very account. If made square and light, they need not be unsightly, and the cost is not very great.

No directions need be given for the construction of

arbors, or those ornamental trellises usually erected in gardens, as their form and arrangement will vary with the taste of the possessor.

A very neat, simple and efficient support for a single vine trained on the spur system is shown in Fig. 43.

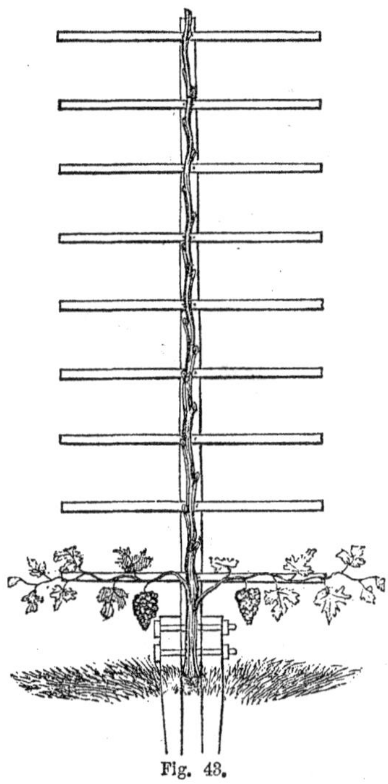

Fig. 43.

It might be constructed so as to be removable when the vine is laid down for the winter. It has even been proposed to have a hinge at the foot of the trellis and lay down trellis and all. But this would be very injudicious.

Stakes are frequently used in vineyards, and also in gardens, but do not present any feature which is not more immediately connected with the subject of training, than with that of the present chapter.

The materials of which trellises, etc., should be made, will as often be governed by local circumstances as by any other consideration. As before stated, cedar, chestnut, locust and oak, are to be preferred for posts, and any tough, light and straight-grained wood for slats. The posts may be charred, where they enter the earth, though we should prefer to soak them for an hour or so in boiling coal tar. This may be readily done in a large pot, or caldron, set up in the field over a temporary furnace. It will of course be wise to give the whole structure a couple of good coats of paint, especially the wire.

The height to which trellises may be carried, depends upon two circumstances—the extent of their shadow and the influence of high training upon the vines. The latter point has been sufficiently discussed, under the head of pruning and training, and we find that at a distance of six feet no ordinary trellis will, in latitudes suited to the culture of the grape, shade its neighbor during the growing season. At other times, shade is not at all injurious. We had prepared extensive tables, giving the distance to which shadows will be thrown by trellises of various heights, at different seasons of the year, and in different latitudes; but omit them for the above reasons.

The only limit which we would set to the height of a trellis, would be our ability to prune the vines and gather the fruit, with the aid of a light stool. Ladders are too unwieldy and involve too much labor, except for a few vines.

CHAPTER X.

PROPAGATION OF THE VINE.

Young vine plants may be raised from seeds, eyes, or cuttings, or by layering or grafting, all which modes are in common practice, though some are only adapted to peculiar circumstances and objects. We shall give a few practical directions for each.

Layering.—This is the mode in which large, thrifty vines may be most rapidly obtained; but it is by no means adapted to general use, where large quantities are required.

To procure a young vine by layering, we take a cane of the preceding year, having a well advanced shoot; and about the middle of June, or first of July, cut it half through, as shown in Fig. 45. It is then bent down and pegged into a hole, about three or four inches deep. It should be well watered, and the application of a little mulch, consisting of long litter, new-mown grass, weeds, or any similar matter, will prove of much service. Roots will soon push, and at the proper time for transplanting, it will have formed a *fine* healthy plant. Larger and stronger vines, which will in some cases bear the succeeding season, may be obtained by layering older and stronger shoots; but the most healthy and, we believe, the most vigorous, plants will be produced by following the directions just given.

Instead of one plant, several may be obtained from

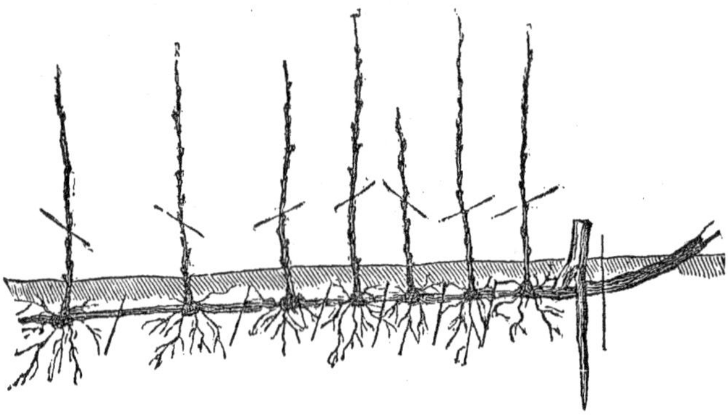

Fig. 44.

the same layer, if it be simply buried its whole length, as in Fig. 44. Roots will start from each joint, and consequently each joint will form a plant. But where a single plant is wanted, the method shown in Fig. 45 will give the finest results.

If very fine plants are wanted, in a short time, the best method is to sink a six-inch (or larger) pot in the gronnd and layer the shoot in it. This is best done by first making a hole in the ground, sufficiently large to receive the pot; then by running the loop end of a doubled cord through the hole in the bottom of the pot, and passing a stick through the loop

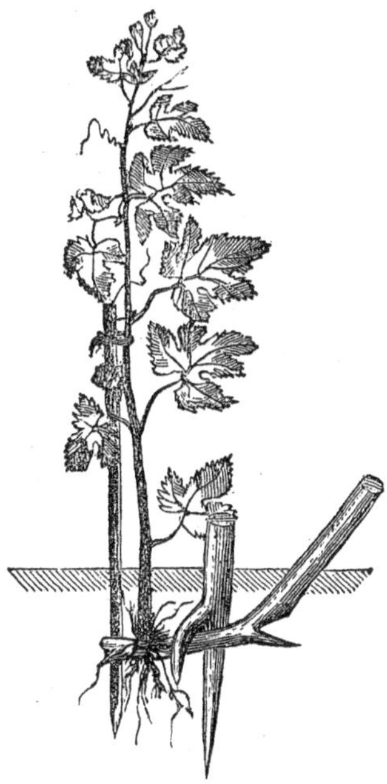

Fig. 45.

or double, it will be easy to tie the shoot in any desired position. The whole process will be readily understood from an inspection of Fig. 46; and we much prefer this plan to pegging down the shoot, or laying on bricks or weights, as pegs are rather uncertain when used in pots, and bricks take up too much room to the prejudice of the roots.

The young plant should be detached from the parent vine in about five or six weeks after layering, and may then be set out in its proper location. As this will be about the latter part of July, or the first of September, the vine will have plenty of time to become well established, and make good roots before winter sets in; and it will form a strong plant, capable of throwing up two permanent canes or producing a specimen bunch of fruit, during the succeeding season—provided, of course, that the variety propagated is of a vigorous and prolific character. An Isabella vine, layered in this manner in an eight-inch pot, threw up, next season, two canes, one twelve and the other sixteen feet. Another, treated in the same way, bore sixteen bunches of fine fruit.

In several instances, we have used common four-inch semi-tubular tile, instead of flowerpots, and with excellent results. They have the advantage of cheapness; but, in other respects, the flower-pot is to be preferred. In some cases, the shoot is drawn through the hole in the bottom of the pot; but although we have tried this in one or two instances, we have not found it either convenient or satisfactory.

A bearing shoot, layered in a good sized pot, or in a common water-pail, may be made to produce a well-rooted plant, which will perfectly ripen several bunches of fruit the same season, even after being removed from the parent plant. This forms a very elegant and ornamental object; but, except as a matter of curiosity, such a process is worthless. In rare instances, perhaps, specimen bunches might be grown

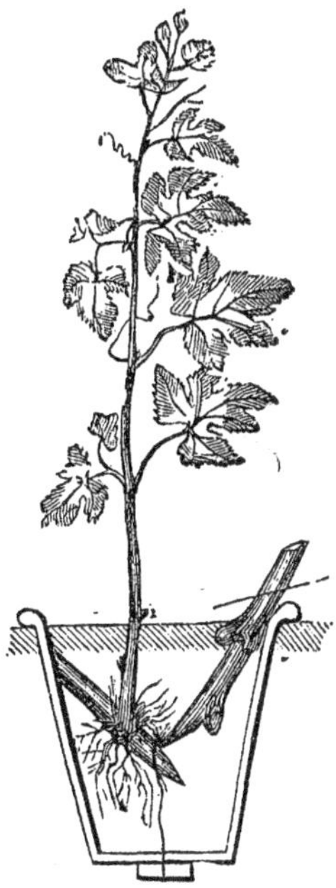

Fig. 46.

out of doors, and perfected in the house, thus avoiding numerous evils to which the finer varieties of the vine are subject in this climate. In all cases, it is essential that the layer be kept moist and warm. As the earth does not seem to be warm enough to induce the formation of roots, before June, it has occurred to us that very early and strong plants might be produced by inserting the pot (Fig. 46) in a slight hot-bed. A few barrow loads of manure would answer every purpose; and by producing roots thus early, strong specimen plants might be procured more easily than by any other method.

It may be well here to state that wood of any age from the oldest gnarled stems to the succulent growth of the current year will root if properly treated. We have always found, however, that the best and healthiest roots always spring from the junction of the old wood with the current year's growth.

It is recommended upon good authority (with which we in general coincide) to separate the plants from the old vine at least by the end of September, unless previously removed. In the case of some varieties, however (Diana, *e. g.*) this will not always answer, as roots are produced with such difficulty that *two* years are often required to make *good* plants.

Cuttings.—Where large quantities of young plants of the common varieties are required, this is one of the cheapest and easiest methods of procuring them. Where wood is plenty, each cutting may consist of several buds or joints, as in Fig. 47. In ordinary cases, however, a length of three buds is sufficient, and we have grown very good plants from cuttings of only one joint in length—that is, having two buds. Indeed, the latter make by far the nicest and cleanest plants, and though not quite so strong at first as those from a greater number of eyes, yet we question if in the long run they would not prove quite their equals.

Fig. 47.

Fig. 48.

Cuttings to be good should be of thrifty, well ripened, close jointed wood—long reedy canes and spindling twigs being alike to be avoided. The best cuttings are those which have the base of the shoot attached, and this may be either as in Fig. 47, where the cutting has been cut away close to the old wood; or it may be a *mallet cutting*, as it is called, where a small section of the two-year-old wood is left, as in Fig. 48.

The proper time for procuring cuttings is at the

regular fall or spring pruning. The cuttings may then be preserved in a cool cellar, either buried in moderately dry sand or simply laid on the floor and covered with straw or leaves. Excessive dryness or dampness are equally to be avoided, and the temperature should be low though never sufficient to freeze the fluids in the cutting. In this state they should be kept until the middle or end of April, as nothing is gained by setting them out at an earlier period. Indeed, we have found those which had been well preserved during the winter and set out in May do quite as well if not better than any others. In some experiments we used the previous year's wood, cut from the vines when the young shoots had grown two inches, and yet in this case they grew finely and made strong plants. But of course this is not an example to be imitated except where it is desired to procure scions of some particular variety, and the opportunity for so doing occurs only at the period indicated. Such cuttings should be set out as soon after being cut off as possible, and if the weather be dry and warm, shading, watching and watering will all be necessary.

In planting cuttings it is best to choose a plot of rather sandy soil (heavy soil will not do); trench it deeply, mixing it with manure *thoroughly rotten and converted into a black mold.* (Any decomposition going on in the soil will ruin the cuttings.) Then plant the cuttings in rows twelve to eighteen inches apart and six to eight inches apart in the rows. They may be planted either in holes made by a dibble or laid in trenches made by the spade—the earth from the next trench being used to fill up the trench in which the cuttings are placed. Some authors direct us to place them perpendicularly, but we have always obtained the best results when they were placed as shown in Fig 49. In placing them, always be careful

Fig. 49.

to have the end bud which is out of the soil uppermost, and be careful that the end be cut with a slant, the same as that in the figure, so that it will not throw the rain *on* to the bud as in that case it may cause it to rot. In long cuttings the upper bud should be left about three inches above the surface of the soil, so that it may not be covered by the coat of mulch, which it is well to apply. Short cuttings must be inserted more deeply, but in all cases the bud should be uncovered unless in very late planting. We are aware that many advise the bud to be covered, but our own experience has been uniformly against it.

Neither should several buds be left above the surface, as they can do no good and require constant watching, as it is important for reasons to be hereafter detailed that only one shoot should be allowed to grow. With cuttings of four or more buds, a very good rule is to place the second bud even with the surface of the ground. In this case the cutting should be turned one-fourth round from its position in Fig. 49 so that the two upper buds may lie one on each side.

Another mode of planting cuttings is to make a hole with a dibble, and after inserting the cutting about two-thirds its length, bend it over and peg it down, as shown in Fig. 50.

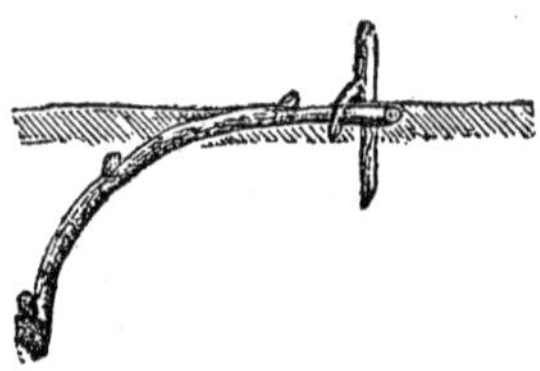

Fig. 50.

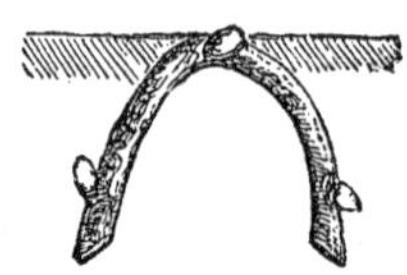

Fig. 51.

Fig 51 shows a method of inserting cuttings which we have practised with success. As it is well known that a cutting will grow, no matter which end is stuck in the ground, a scion is taken containing at least three buds, and after bending it into a semicircular form, both ends are inserted in the ground, leaving the middle bud above the surface. As soon as this bud begins to grow, it will be supplied with nutriment from both ends and will make rapid progress. The plants produced by this method are very strong and if designed to remain where they are first set out they give very certain and satisfactory results. But they do not transplant well.

The following plan described in Miller's "Gardener's Dictionary" is one which we have tried with success: "Having an Iron bar of an Inch or more in Diameter, a little pointed at the End, they therewith make a Hole directly down about three Feet and a Half deep; then, being provided with an Instrument they call a Crucciala, having a Handle of Wood like that of a large Auger and the Body of Iron four Feet long and more than half an Inch in Diameter, at the End of which there is a Nich something like a half moon, they after twisting the End of the Cutting, put it therein, and force it down the Bottom of the Hole, where they then leave it, and afterward fill up the Vacancy with fine sifted Earth or Sand; observing to tread the Earth close to the Plants, which otherwise (unless it be stiff Land) is often inclinable to be Loose and Dry, especially if Rain does not soon follow their Planting; and it is incredible how many Vines three Persons can in this Manner plant in one Day, viz., upward of two thousand."

In our own practice we simply make a deep hole, insert the cutting, fill up with *dry* sand and give a liberal supply of water. By inserting the cutting as deeply as here advised it is placed beyond reach of drought, though the lower portion rarely throws out roots unless the soil be very favorable.

EYES.—Where the aid of a hot bed or propagating house can be obtained, eyes afford an easy and rapid mode of multiplying vines. They are usually cut about two inches long, containing only one bud, and are started in February or March in pots or boxes filled with a mixture of sand, leaf-mold, and soil. The buds are either buried half an inch deep or placed even with the surface of the earth, according

Fig. 52. Fig. 53.

to the ideas of the operator, and generally form strong plants, which may be set out in the open ground in June or July. Figs. 52 and 53 illustrate the position of the cutting in relation to the soil. This method of propagating requires skill and great care and is seldom employed except by professional men.

GRAFTING is seldom employed in the case of those common varieties which are easily raised from cuttings or eyes. Diana and Delaware are, however, sometimes raised by this method, and to the amateur it is one of the most important processes, as by this means a new and rare variety can be fruited some years before a young plant would come into bearing. Loudon gives the following directions for performing the operation:

"Cleft grafting the vine is shown in Fig. 54, in which *a* is a bud on the scion, and *b*, on the stock,

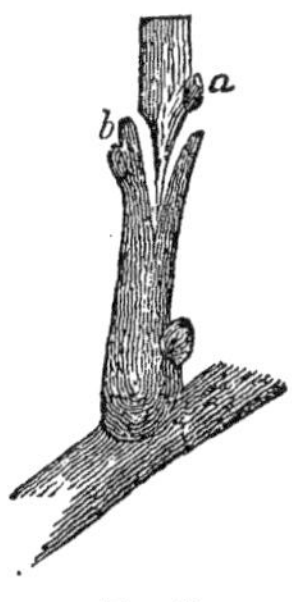

Fig. 54.

both in the most favorable position for success. The graft is tied and clayed in the usual manner, excepting that only a small hole is left in the clay opposite the eye of the scion, for its development. In grafting the vine in this manner when the bud (*b*) on the stock is developed, it is allowed to grow for ten or fourteen days, after which it is cut off, leaving only one bud and one leaf near its base to draw up sap to the scion till it be fairly united to the stock. The time of grafting is when the stock is about to break into leaf, or when it has made shoots with four or five leaves. By this time the sap has begun to flow freely, so that there is no danger of the stock suffering from bleeding; though, if vines are in good health and the wood thoroughly ripened, all the bleeding that takes place does little injury."

Root-grafting the vine is also frequently practised. For this purpose, saddle-grafting is most suitable, the scion being properly secured to the stock with waxed cloth or paper.

Lindley, in the "Gardener's Chronicle," states that the great secret of success in grafting the vine is to keep the scion dormant until the stock has so far developed its leaves and shoots as to be beyond the reach of danger from bleeding. His directions are as follows:

"Shorten the branch or shoot at the winter pruning, to the most eligible place for inserting the graft. The graft should be kept in sufficiently moist soil till the time of performing the operation, and for a week previous in the same temperature as that in which the vines to be operated upon are growing. When such portions of the latter as are shortened for receiving the grafts have made a bit of shoot, graft as you would other fruit-trees, taking care, however, to preserve the shoot at the top in claying, and till the buds on the scions have pushed, then shorten it back. Inarching may be performed at any time after the vines have started, so far as not to bleed."

Speechly, however, names the middle of March as the best time for grafting the vine in the open air; and his directions are so lucid that we offer no apology for quoting them:

"In general, vines should be grafted about three weeks before they begin to break into bud.

"Upon small stocks, not more than one inch in diameter, cleft grafting will be found the most proper; but upon larger stocks, whip grafting is to be preferred.

"In both methods, much care should be taken in fitting the scion and stock together, and the operation should be performed with great exactness.

"When the stock and scion are well fitted the graft should be fastened with the strands of bass matting, and should then be covered with clay in the usual way.

"Vines do not harmonize with so much freedom as

commoner fruit; for though the scion will sometimes begin to push in a few weeks, yet it will frequently remain in a dormant state for two or three months; and during this period it will be necessary to strip the stock of all the shoots it may produce, as soon as they appear; and, in order to preserve the scion in a vegetative state, it will be necessary to keep the clay moderately moist, which may easily be effected by wrapping it round with moistened moss, and keeping the moss constantly sprinkled with water.

"When the scion has made shoots five or six inches long, the clay and bandage should be carefully taken off; and the clay may be removed without injuring the graft, when it is in a moist state.

"Vines will frequently prove successful by both the above-mentioned methods, but still the most eligible way of all, seems to me, to be that of grafting by approach. Indeed, I have seldom known any plants miscarry, that have been grafted in this way. Now in this case, it is necessary to have the plant, intended to be propagated, growing in a pot. Strong plants that have been two or three years in pots are to be preferred, but plants from the nursery may be potted and grafted in the same season, if brought into a hot-house or vinery; for the great warmth of either will generally cause plants, brought out of the open air, to push with vigor, and to form new roots, which will support the plant, and greatly facilitate its forming a union with the stock.

"I have constantly had fine grapes, and the grafts have made good wood, the first season, by every method of grafting, but particularly by the last. In which it is obvious that the graft has a double support, viz.: from the stock, as well as from the plant in the pot.

"In this method it will be necessary to let the clay and bandage remain two or three months after the graft has formed a union; for if taken off at an earlier period, the grafted part of the plant will be very liable to spring from the stock.

"The pot should be plentifully supplied with water till the month of August, when the graft should be separated from the plant in the pot. Two or three inches of wood below the bottom of the graft may be left, but should be taken clean off at the next winter's pruning."

Grafting is a common practice in the vineyards of France. Chaptal's directions are as follows:

"Having selected a healthy stock, it is, just when the sap is beginning to flow, taken off with a clean cut an inch or two below the surface of the ground. The upper portion of the stock, which must be perfectly free from knots, is split evenly down the centre and pared quite smooth within, of a sufficient size for the reception of the scion. The latter is pruned to three eyes in length, having the lower part cut in the form of a wedge, commencing about an inch beneath the lowest eye, and gradually tapering to the bottom. It is then inserted as far as the lowest bud into the cleft of the stock: the second bud is level with the surface of the ground, which is drawn close around it, and the uppermost is quite above the soil. Great

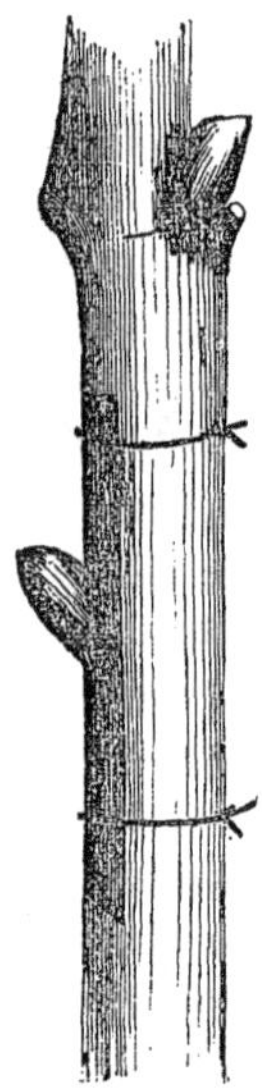

Fig. 55.

care is necessary in adjusting the scion, that its bark may touch that of the stock in every possible point.

"The whole is then bound round with a pliable osier

which retains the scion in its proper place. The best season for grafting the vine is just when the warmth of spring sets the sap in motion, and it should be performed when the sky is cloudy, with the wind blowing from the southeast or southwest. Whenever a northerly wind or great drought prevails, it is better to delay the operation; a burning sun or cold wind would arrest the course of the sap by drying up the vessels at the point of union. Neither is it advisable to graft in rainy weather, because the water will trickle down into the incision, and prevent the union between the scion and stock. The best time for taking off the grafts is in a dry day toward the end of autumn, when the sap is still. They should be cut off with a portion of the old wood adhering, which will assist in preserving them until wanted for use. They should be plunged two or three inches deep in damp sand, and kept in a cool cellar, where neither heat nor frost can penetrate. Twenty-four hours previously to being used, they should be taken up, and that part which had before been in the sand should be laid in water."

He, moreover, states that "the vine is thus grafted with so much facility, and the union between scion and stock is so perfect, that no plant appears more adapted for this mode of propagation."

We have never met with an example of budding

9

as practised on the grapevine. The following process, which is described as budding in the "Gardener's Chronicle" for 1844, is in reality a species of grafting:

"Bud about the first week in March, or as soon as the sap begins to rise. Cut an eye about three inches in length, having attached as much wood as you can get with it; at each end of the eye cut off about a quarter of an inch of the upper bark, making the ends quite thin. Next measure off the exact length of the bud on the bark of the vine intended to be budded, and make a niche slanting upward at the upper part; and another slanting downward at the bottom. Then take the piece neatly out, so that the bud may fit nicely in, and by making the niche, as stated above, each end of the bud is covered by the bark of the shoot. Bind the bud firmly round with matting, and clay it, taking care, however, that the clay does not cover the eye of the bud. Then tie it round with moss, and keep it constantly damp, and as the sap rises in the vine the bud begins to swell. When the vine commences to push out young shoots, take the top ones off, in order to throw a little more sap into the bud, and as you perceive it getting stronger take off more young shoots, and so continue until you have taken off all the young shoots. Budding can only be performed where the long-rod system is practised, as in that case you have the power of confining the sap to the bud, which will grow vigorously. As soon as you perceive this, cut the vine down to the bud. Budding has the advantage over grafting of not leaving an unsightly appearance where the bud was inserted. A bud likewise grows more luxuriantly. Allow the matting to remain until about the month of September."

Mr. Knight was accustomed on some occasions to employ two distinct ligatures to hold the bud of his peach-trees in its place. One was placed above the bud inserted, and upon the transverse section through the bark; the other, which had no further office than that of securing the bud, was employed in the usual way. As soon as the bud had attached itself, the ligature last applied was taken off; but the other was suffered to remain. The passage of the sap upward was in consequence much obstructed, and buds inserted in June began to vegetate strongly in July: when these had afforded shoots about four inches long the remaining ligature was taken off to permit the excess of sap to pass on; and the young shoots were nailed to the wall. Being there properly exposed to light, their wood ripened well, and afforded blossoms in the succeeding spring.

Might not the principles here indicated be applied with advantage to the foregoing method of budding (grafting?) the vine?

A method which partakes partly of grafting and partly of inarching is shown in the annexed figure. Fig 56. Here the graft is covered with soil which

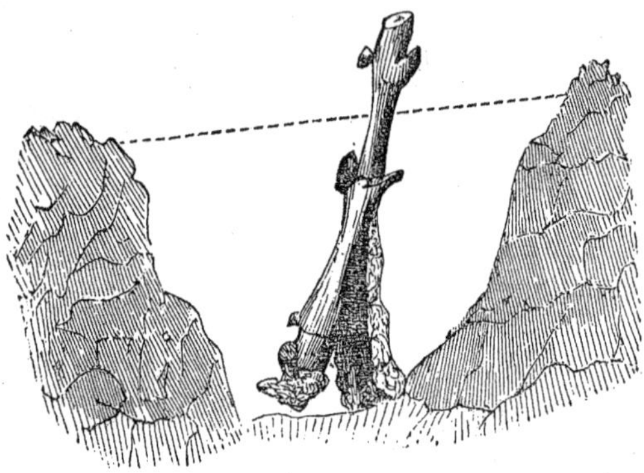

Fig. 56.

supports it in the same manner as a cutting, while at the same time it receives nutriment from the stock.

Seed Hybridization.—Young vines are never raised from seed, except for the production of new varieties; but, as this subject is deservedly attracting very general attention at present, a few practical hints thereon may prove acceptable.

Ever since Bacon observed that "The compounding and mixture of plants is not found out, which, nevertheless, if it be possible, is more at command than that of living creatures, wherefore it were one of the most noble experiments touching plants to find this out; for so you may have a great variety of plants and flowers yet unknown. Grafting doth it not; that mendeth the fruit, or doubleth the flower, but it hath not the power to make a new kind," it has been the constant endeavor of good gardeners to improve the qualities of domestic plants by judicious mixtures of varieties.

Bradley, we believe, was the first who undertook to produce hybrid plants; but since his day, it has been attempted by almost every celebrated horticulturist.

The limits of hybridization amongst plants have never been thoroughly ascertained, although it is a subject of deep importance. For in the animal kingdom we know that while cross breeding (or intermixture of varieties) has been productive of the best results, hybridization, or muling has been successful in but very few instances, at least so far as practical good is concerned. Now whether the different species of the vine, as *vitis vinifera*, *vitis labrusca*, *vitis cordifolia*, etc. are so far removed from each other as to produce mules by their intermixture, or whether they are varieties and will freely cross-breed, has not yet been fully determined.

To examine this subject, however, with sufficient fullness to be useful would far exceed our limits.

Neither can we enter upon a discussion of the claims of the rival theory of Van Mons. We shall therefore rest content with a few practical directions.

Speechly, who was probably the first to attempt the improvement of vines by cross-breeding, directs us to bring the flowering branches of the two kinds of vines into close proximity—they being, of course, in the same stage of maturity.

No very superior varieties followed his attempts, however; and this is not to be wondered at, as it is probable that branches might be entwined a thousand times without effecting the result aimed at; for no means are taken to bring the pollen of the different flowers into contact with the stigma of the others, and although we have no reasons to doubt the possibility of super-fœtation, (seeing it is well-known to occur in the higher animals) yet no means are here taken to produce even this.

J. Fiske Allen, one of our most successful grape growers and the originator of some new and excellent varieties, gives the following directions upon this point.

"The applying the pollen, or farina of one variety to the pistil, or stigma of another, is the surer method of proceeding to obtain new sorts in the shortest time; and this is called hybridizing.

To do this properly, the bunch should be thinned

of three-quarters of the buds; the lower part should be cut away entirely (immediately before inflorescence), the strongest buds always being left.

Observe them closely, and as soon as the flowers open, with sharp scissors clip the *anthers*, being careful not to injure the pistil; with a soft brush, apply the pollen from the kind to be used in impregnation; or the whole bunch which is to furnish the pollen may be cut from the vine and gently rubbed or applied to the bunch, by frequently striking them together on every side. This should be repeated several days, until it is evident that the fruit is all impregnated; a fresh bunch with the pollen in a suitable condition, must be had at each operation.

"The pollen must be dry and in a falling condition, to be fit for the purpose. If your vines are so situated that a branch to be acted upon can be brought into contact with the branch of another kind, and the bunches interlaced, this will be a good method of proceeding—cutting away the males part of the blossom from the kind that is to ripen the seed for the new kinds."

Fig. 57.

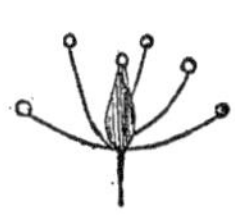
Fig. 58.

Fig. 59.

"Fig. 57 is a magnified representation of the bud

of the grape. Figs. 58 and 59 show the blossom. The change from the bud to the blossom is usually rapid, and takes place about thirty to forty days after the shoot appears in the spring which bears the fruit. This bud, which forms the blossom, consists of a covering, or cap, and the embryo berry with five anthers, which, when the time for inflorescence has come, is raised, or lifted, by the anthers, and the wind blows this cap free.

"The third is the blossom or embryo grape, with the anthers clipped and deprived of their farina; on the top of the embryo is the pistil; upon this is to be placed the farina, or pollen of the male plant; when this is done, impregnation takes place, and the embryo rapidly swells off. If the operation has not been effectual, the berry will remain as it is. When the grape has attained one-third or one-half its size, it remains stationary two or three weeks, and at this, time it is perfecting the seed. When this is done, the fruit begins growing again; thus it appears the seed will vegetate, even if the fruit does not ripen sufficiently to be eatable."

Loudon's directions for saving and sowing seed are as follows: "Grapes for seed should be permitted to remain on the plant till the fruit is perfectly mature, and the seeds are of a very dark brown color. They should be separated from the pulp, and preserved till

February or the beginning of March. They should then be sown in pots filled with light fresh mold, and plunged in a moderately warm hot-bed; they will come up in from four to six weeks, and when the plants are about six inches high, they should be transplanted singly, into forty-eights, and afterward into pots of a larger size. Water gently, as circumstances require, allow abundance of light and air, and carefully avoid injuring any of the leaves. Cut down the plants every autumn to two good buds, and suffer only one of these to extend itself in the following spring. Shift into larger pots as occasion requires, till they have produced fruit. This, under good management, will take place in the fourth or fifth year, when the approved sorts should be selected, and the rest destroyed, or used as stocks on which to graft or inarch good sorts."

CHAPTER XI.

MANURE.

Manure* may be defined to be anything added to the soil to increase its fertility, whether by mechanical or chemical action. Substances serving the first purpose have been alluded to under the head of soils. The latter will now occupy our attention.

In a former chapter sufficient practical directions were given for the preparatory enrichment of the soil and for the annual top-dressing of the borders; it will now be our object to consider in detail the character of the nutriment required by the vine; the sources whence it may be derived; the various modes of its application; and its effects upon the plant.

It is a well-established fact, that unless the soil in which any plant is placed contains all the elements necessary to the formation of such plant, no healthy growth can ensue. Hence our first step must be to inquire into the chemical constitution of the grapevine, or at least of its ashes, those elements which

* QUERY.—To what extent was Jethro Tull's idea of horse-hoeing, as a substitute for manure, anticipated by those who first used the word *manure* (*manœuvrer*—to work with the hand), before it was employed to express the addition of matter to the soil, with a view to increase its fertility? One old English author speaks of the Commonwealth of England as being "gouerned, administered, and manured by three sorts of persons," etc.

are dissipated during combustion being abundantly supplied from the atmosphere.

The following are a few of the most reliable analyses which have been published:

Dr. Emmons found the wood of *Vitis Labrusca* (Isabella?) to contain: Water, 40.26; dry matter, 59.74; ash, .98.

Full-sized leaves of Catawba, picked June 2d: Water, 72.388; dry matter, 27.612; ash, 2.138. Per centage of ash calculated on dry matter, 7.746.

Leaves of Catawba grape picked June 2d. Analysis of ash:

Carbonic acid	3.050	Potassa	13.394
Silicic acid	29.650	Soda	9.698
Sulphuric acid	2.062	Chlorine	0.741
Phosphates	32.950	Organic acid	2.250
Lime	4.391		
Magnesia	1.740		96.926

An analysis of wood and bark of wild vine gave

	Wood.	Bark.
Potassa	20.84	1.77
Soda	2.06	9.27
Chlorine	0.02	0.40
Sulphuric acid	0.23	trace
Phosphate of lime	15.40	5.04
Phosphate of peroxide of iron	1.20	5.04
Carbonic acid	54.83	32.22
Lime	17.33	39.32
Magnesia	4.40	0.80
Silex	2.80	14.00
Soluble silica	0.00	30.00
Coal and organic matter	2.20	1.70
	100.21	100.86

The following tabulated analyses by Crasso & Walz explain themselves:

ANALYSES OF ASH OF GRAPE VINE.

COMPONENT PARTS.	CRASSO.									WALZ.		
	Unripe Blue Grapes—Ash of Juice.	Ripe Blue Grapes—Ash of Juice.	Ripe Blue Grapes—Ash of Juice.	Ripe Green Grapes—Ash of Juice.	Blue Grapes—Ash of Skin.	Green Grapes—Ash of Skin.	Blue Grapes—Ash of Seed.	Green Grapes—Ash of Seed.	Small Burgundy Vines—Ash of Wood.	Cleven Vines, with leaves.	Reisling Vines, with leaves.	Drollingen Vines, with leaves.
Potassa	66.33	65.04	71.85	62.74	41.65	46.89	27.87	29.45	44.15	7.04	6.31	11.96
Soda	0.33	0.42	1.20	2.66	2.13	1.62	......	......	3.45	9.20	7.13	3.41
Lime	5.20	3.37	3.39	5.11	20.31	21.73	32.18	35.57	36.04	21.72	24.54	31 07
Magnesia	3.27	4.74	3.97	3.95	6.02	4.45	8.53	8.59	4.77	6.81	4.76	9.22
Sesquioxide of Iron	0.73	0.43	0.09	0.40	2.11	1.97	0.45	0.65	0.54	......	......	.....
Proto-sesquiox. of Manganese	0.82	0.75	0.10	0.30	0.76	0.51	0.35	0.45	0.11	0.03	0.13	0.17
Sulphuric Acid	......	5.54	3.65	4.89	3.48	3.88	2.40	2.61	1.82	1.27	1.36	1.13
Hydrochloric Acid	5.19	......	......	......	......	......	......	......	......	......	......	.....
Chlorine	0.70	1.03	0.47	0.70	0.49	0.71	0.27	0.35	0.85	......	......	.....
Silicic Acid	1.99	2.10	1.19	2.18	3.46	2.57	0.95	1.27	1.22	3.54	3.41	3.25
Phosphoric Acid	15.38	16.59	14.07	17.04	19.57	15.66	27.00	21.05	7.05	......	......	.....
Phosphate of Alumina	......	......	......	......	......	......	......	......	......	1.15	0.91	1.00
Phosphate of Sesquiox. of Iron	......	......	......	......	......	......	......	......	......	2.21	2.38	2.47
Phosphate of Lime	......	......	......	......	......	......	......	......	......	11.17	19.63	9.25
Chloride of Sodium	......	......	......	......	......	......	......	......	......	1.27	1.73	1.65
Charcoal, Sand, CO^2 and Loss	.....	......	......	......	......	......	......	......	......	34.56	27.68	25.30
Total	100.00	100.00	100.00	100.00	100.00	100.00	100.00	100.00	100.00	100.00	100.00	100.00
Percentage of Ash	0.259	0.340	0.409	0.290	3.745	4.321	2.776	2.837	3.692	6.19	7.74	6.00

Such being the normal constituents of the vine and of its fruit, and the latter being, in almost all cases, removed from the soil in which it was produced, it is obvious that a process of exhaustion must be constantly carried on, which, if not counteracted, must, in a short time, perceptibly reduce the crop.

The means by which the matter thus removed from the soil is restored, are of three kinds: First, the action of the plants themselves, or of man upon the subsoil; secondly, rain; and thirdly, by the direct addition of the requisite elements, through the agency of man and animals.

Although the soil has, to a certain extent, the power of separating salts and gases from the water which passes through it, the drainage water still retains a certain proportion of valuable matter,* and consequently the subsoil also becomes saturated to a greater or less extent with these same elements. Hence one of the effects of trenching is not only to bring up unexhausted soil to the surface, but to return those matters which had previously been washed out of the upper soil by the rains. The plants themselves occasionally bring up some of this matter, sending

* A series of valuable analyses and experiments upon this point appeared lately in the transactions of the Highland (Scottish) Agricultural Society, which the reader who desires to pursue this subject would do well to consult.

down roots deep into the subsoil if it is open and porous.

Rain is another important source not only of ammonia and gases, but of mineral matter. We quote the following from Lindley's "Theory of Horticulture:"

"The researches of chemists have shown that all rain water contains ammonia, a compound of hydrogen and nitrogen, and thus the source of the nitrogen absorbed by plants was explained. But it has also been shown, especially by M. Barral, that other substances upon which plants feed are contained in rain water to a much greater amount than was suspected. This observer was led, during six months of 1851, to examine minutely the water collected in the rain gauges of the Observatory of Paris. His mode of investigation is declared by Messrs. Dumas, Boussingault, Gasparin, Régnault, and Arago, names foremost in French science, to be free from all objection, and to bear the most counter trials to which they could expose it. M. Barral states, that although the quantities of the following substances varied in different months, yet the monthly average from July to December, inclusive, was as follows:

"SUBSTANCES IN A CUBIC METRE OF RAIN WATER.

	GRAMMES.		GRAINS.
Nitrogen, ...	8.36	=	129.
Nitric Acid, ...	19.09	=	294.

	GRAMMES.		GRAINS.
Ammonia,	3.61	=	55.7
Chlorine,	2.27	=	35.
Lime,	9.48	=	100.
Magnesia,	2.12	=	32.7

"He did not ascertain whether all these substances are contained in rain water collected at a distance from towns. But Dr. Bence Jones found at least nitric acid in rain water collected in London, at Kingston in Surrey, at Melbury in Dorsetshire, and far from any town at Clonakelty, in Ireland. If we assume that M. Barral's averages represent what occurs on an English acre, the quantity of such substances deposited on that extent of ground may be safely estimated as follows:

"The average depth of rain which falls in the neighborhood of London is well ascertained to be about twenty-four inches per annum. This is at the rate of 87,120 cubic feet, or 2,466 cubic metres of rain water per acre; and this, according to the proportions per cubic metre in the preceding table, would afford anually of—

Nitrogen,	45½	lbs.
Nitric acid,	103	"
Ammonia,	19½	"
Chlorine,	12½	"
Lime,	35	"
Magnesia.	11	"
Amount total per acre,	227	"

"Of these substances, the three first are of the utmost importance, on account of their entering so largely into the indispensable constituents of the food by which vegetable life is sustained. The quantity of ammonia thus ascertained to exist, is about what is expected in two hundred weight of Peruvian guano; and bountiful nature gives us, moreover, nearly one hundred and fifty pounds of nitrogenous matter equally suited to the nutrition of our crops."

But although nature is thus liberal in supplying the necessary wants of her children, man desires returns rather more extensive than is merely necessary for the good of the plant. He therefore adds directly to the soil those matters which contain proper nutriment for the vine. In doing this, however, it is not neces sary to follow very accurately any recipe founded upon the analysis of the vine, provided we obtain sufficient of those elements which are most wanted. If we only spread a liberal table, the vine will select its own food.

Of all applications to the soil, none deserve more confidence than well rotted barnyard manure; from time immemorial it has been the staple reliance of the gardener and farmer and few are the instances in which its judicious application has been known to fail.

That it may do good and not harm, however, *when*

mixed with the soil in which plants are growing, it is necessary it be thoroughly rotten. However much may have been written about the waste incurred by allowing manure to decompose, it is a well know fact that thoroughly decomposed manure is beneficial to most plants, while decomposing or fermenting manure is frequently prejudicial. This probably arises from the fact that all bodies while undergoing decomposition exert a catalytic action on any organized matter in contact with them. Thus decomposing manure directly tends to produce rot in the roots or other parts of plants with which it comes in contact.

The proper time for the application of manure to a vine border has been a subject of much discussion. Our plan is to apply it as a top-dressing in the fall and fork it in in the spring. It thus serves to keep the border warm and the soluble portions are washed down amongst the roots of the vine by the winter snow and rain, thus reaching it in a most effectual manner. To assist this process, the border should be loosened with a fork before the manure is laid on.

Of all the substances entering into the composition of a manure heap none have a better influence upon vines than bones. In the formation of a border they are of essential utility, affording for a long period a constant source of nutriment. The avidity with which the roots of the vine seek such a depot of food may be easily seen by placing a large porous bone amongst the roots of a vine. In a few months it will be literally covered with rootlets which have sought it out and find their nutriment in its recesses.

Leather, hair, horns, hoofs, woollen rags and other animal offal possess a similar action to bones. They all possess the valuable property of lying undecomposed in the soil for long periods, yet yield readily to the disintegrating action of plant roots. Hence, while they afford abundant and valuable nourishment to the vine, it is not surfeited by them as this nutriment must be wanted and sought before it will be given up.

Ashes of wood, whether fresh or leached are a powerful manure for the vine, and probably contain all that it requires. Leached ashes may be applied as a top dressing in almost any quantity with excellent effect, but a more cautious use must be made of fresh wood ashes, they being much more powerful and caustic. Coal ashes have hitherto been deemed utterly worthless, and are usually thrown into the street. To some soils, however, particularly those which are too heavy, they are a very useful addition, and as they are a powerful absorbent, there is no doubt that if mixed with night soil, or some similar matter, they would prove an excellent article—more lasting, and consequently better than night soil by itself. They should never be thrown away, however, as they contain lime, iron and minute, though appreciable quantities of alkalies, soda usually predominating. It is also quite possible that they contain minute traces of phosphates, though in no analysis with which I have met is it mentioned. Where the coal has been burned at a high temperature the alkali is in general reduced, and the metal volatilized.

The dung and urine of animals forms a powerful manure. The solid excrements of all these are best mixed with some absorbent, as plaster, charcoal, burnt clay, etc.; or thoroughly decomposed in contact with vegetable matter, as straw, leaves, etc. The liquid and soluble portion may be used as liquid manure, or may be poured over the fermenting dung heap.

The dung of birds, as hens, pigeons, etc., and also guano, form a very convenient and most excellent top dressing for vine borders, but are better when applied as liquid manure during the growing season.

An excellent manure may be made as follows: Sink a hole in any convenient part of the premises and fill up with saw-dust. On this pour all the urine that can be obtained from time to time, and keep closely covered with a broad cover.

When sufficient has been added, or when the smell becomes offensive, remove the cover and place a pile of charcoal, burnt clay, coal ashes, or other absorbent on top of it, and allow it to lie for a few weeks. At the end of that time, a mass of matter will have been produced almost equal to guano.

Road scrapings form a good top dressing for most soils. Hoare considers them unrivalled for the grape vine, and such was the opinion of Speechly, who tells us: "The dust, or dirt, from roads, consists principally of the following particulars: first, the soil of the vicinity; secondly, the dung and urine of horses and other animals; and thirdly, the materials of the road itself when pulverized. Various other matters may be brought by winds, and by other means, but the foregoing may be deemed the principal. The first of the above articles is brought to roads by the wheels of carriages, and the legs of horses and other animals; the last is the worst part of the materials, as the dust and scrapings of roads, made and mended with soft stone that grinds fast away, is much inferior in its vegetating quality to that which is collected from hard roads. On the whole, however, this ingredient of compost from the roads is unquestionably in general of a fertile nature, which may be attributed in part to the dung, urine, and other rich materials, of which it is composed, and in part to a kind of magnetic power, impressed upon it by friction and its perpetual pulverization.

"The nature of this road earth ought to be duly considered, when used in the vine compost, and its proportion adjusted according to its quality. In a sandy country it will naturally abound with particles of sand, and long and continued rains will, of course, wash away *its* best parts. High winds, too, in dry weather, will as certainly deprive it of its lightest and finest parts, especially when roads lie on eminences, or enjoy an open exposure. Those materials from roads are therefore preferable, which are produced from an inclosed track in a low situation Pavements, however, and hard roads, produce the best sulture of all. The compost is much better when collected in a moderate dry state, than when it is either very wet or dusty. If scraped off the road in a wet and soft state, when it is become dry it will be hard and cloddy, and will require time to bring it to a proper condition.

"When thus circumstanced, the best way of recovering it is to give it frequent turnings in hard, frosty weather."

Dead animals may be used in a vine border if placed sufficiently far from the roots to allow of their being decomposed, before the roots reach them, as previously remarked.

Any *decaying* matter in a border is very detrimental.

Charcoal is one of the best additions to any soil. It should be well burnt, however, and free from all smell of creosote, as this substance is rather prejudicial to the roots of the grape vine, although it seems to agree with some plants; (chiefly alliaceous, for which soot is a specific). On this account, when used for drainage in pots, it should be reburnt.

I am informed by one successful grape culturist that unless this precaution of reburning is taken with most charcoal, it is rather prejudicial to the roots of young vines in pots than otherwise. I have found, however, that if well sprinkled (it need not be saturated) with putrid urine and allowed to lie for some time, it loses its injurious qualities and retains abundant nourishment, which is gradually given off to the roots of the plants as they require it. When used as a compost for enriching a vine border it had always better be saturated with night soil or urine. Even brick rubbish, if so treated, becomes of great value.

Most of these solid matters are best added to the soil in the original formation of the border. This is especially the case with the prunings of the vine, than which nothing can be more valuable. If added when the border is first formed, it will not only furnish nutriment for the vine, but will tend to keep the soil open and porous. For our established vines, therefore, it will be best to depend upon liquid manure and autumn top dressing of stable manure, and all solid matters may go to the formation of new vineyards, of which we suppose there will in general be an annual addition. But where no new borders are being formed, it would be well to open trenches between the rows of vines, in which such matters might be buried. If this were done immediately after the vintage, the roots would recover the same season from any wounds they might receive, and the ultimate gain would greatly overbalance any temporary injury. In doing this, it will of course be best to enrich but a small extent of border each year and do it thoroughly, so that it may afford a supply during many succeeding seasons.

LIQUID MANURE.—Of all the forms in which manure can be applied, the liquid manure is the most convenient and the most effective. No garden or vineyard should be without a tank of this article, as its judicious application will often enable us to mature a fine crop under very unfavorable circumstances, its great advantage consisting in the immediate results obtained; though this very quality, renders it a dangerous article in the hands of those who do not thoroughly understand its proper application. To prepare and preserve liquid manure, two tanks with good covers should be made in some convenient spot. In small gardens, barrels, such as are used for hydraulic cement, will answer—larger establishments, of course, requiring something more capacious. They should be filled with chamber and kitchen slops and soap suds, the latter being generally added warm. On the large scale, when horse, cow, and other manure can be obtained, it may be mixed with water and added to the contents of the barrels. Hen manure is one of the most valuable additions. *Two* barrels should be used, so as constantly to have some of the manure *thoroughly* decomposed.

After standing for a week or ten days, it will be fit for use, and may either be applied to the surface of the border, or what is far better, introduced by means of subterranean drains or channels. These may consist simply of long wooden boxes, bored full of small holes and sunk about twelve inches beneath the surface, or of common horse shoe tiles placed in a similar position. Under any circumstances, it must have a tube at one end rising up to the surface, through which the liquid may be poured and which may be closed on the approach of winter so as to exclude frost. In very small gardens, it may prove sufficient to sink one or two flower pots in the border. These, being filled with the liquid manure, it will soak down amongst the roots without the possibility of loss by evaporation from the surface of the ground. The pots should, of course, ordinarily be kept covered.

Liquid manure is such a powerful agent, that there is only one season of the year at which it can be applied; that is from the time the first leaves are well developed until the fruit is fully formed. During this period a *very weak* solution may be applied in *large quantities* once or twice a week. The culturist, however, must remember that the solution must be *weak*—say one pailful of the contents of the barrels to six or ten pails of water, according to the strength of the original liquid.

To prepare extemporaneous liquid manure ready for application to the borders or drains, dissolve two or three ounces of guano in a gallon of rain water, and allow to stand some time, stirring occasionally.

The principles which govern the application of this useful and powerful agent, are so clearly set forth by Dr. Lindley, in the last edition of his "Theory of Horticulture," that we cannot do better than quote from him.

"In order that the full effects of liquid manure should be felt without injury, it is indispensable: 1, that it should be weak, and frequently applied; 2, that it should be perfectly clear; 3, that it should be administered when plants are in full growth. If strong, it is apt to produce great injury, because of the facility with which it is absorbed, beyond the decomposing and assimilating power of plants. If turbid, it carries with it in suspension a large quantity of fine sedimentary matter, which fills up the interstices of the soil, or, deposited upon the roots themselves, greatly impedes their power of absorption. If applied when plants are torpid, it either acts as in the case of being over strong, or it actually corrodes the tissues.

"Let the manure be extremely weak; it owes its value to matter that may be applied with considerable latitude; for they are not absolute poisons, like arsenic and corrosive sublimate, but only become dangerous when in a state of concentration. Gas water illustrates this; pour it over the plant in the caustic state in which it comes from the gas-works, and it takes off every leaf, if nothing worse ensues. Mix it with half water—still it burns; double the quantity once more—it may still burn, or discolor foliage somewhat. But add a tumbler of gas water to a bucketful of pure water, no injury whatever ensues; add two tumblers full, and still the effect is salubrious, not injurious. Hence it appears to be immaterial whether the proportion is the hundredth or two hundredth of the fertilizing material.

"Manuring is, in fact, a rude operation in which considerable latitude is allowable. The danger of error lies on the side of strength, not of weakness.

"To use liquid manure very weak and very often is, in fact, to imitate nature, than whom we cannot take a safer guide. This is shown by the carbonate of ammonia, carried to plants in rain, which is not understood to contain, under ordinary circumstances, more than one grain of ammonia in 1 lb. of water; so that in order to form a liquid manure of the strength of rain water, 1 lb. carbonate of ammonia would have to be diluted with about 7,000 lbs. weight of water, or more than three tons. Complaints which have been made of guano water and the like are unquestionably referable to their having been used too strong.

"It must be borne in mind: 1, That liquid manure is an agent ready for immediate use, its main value depending upon that quality; 2, that its effect is to produce exuberant growth; and 3, that it will continue to do so as long as the temperature and light required for its action are sufficient.

"These three propositions, rightly understood, point to the true principles of applying it; and if they are kept in view, no mistakes can well be made.

"With fruit, the period of application should be when the fruit, not the flowers, is beginning to swell. Nothing is gained by influencing the size or color of the flower of a fruit tree; what we want is to increase the size or the abundance of the fruit. If liquid manure is applied to a plant when the flowers are growing, the vigor which it communicates to them must also be communicated to the leaves; but when leaves are growing unusually fast, there is sometimes a danger that they may rob the branches of the sap required for the nutrition of the fruit; and if that happens, the latter falls off. There, then, is a source of danger which must not be lost sight of. No doubt the proper time for using liquid manure is when the fruit is beginning to swell, and has acquired, by its own green surface, a power of suction capable of opposing that of the leaves.

"At that time liquid manure may be applied freely, and continued from time to time as long as the fruit is growing. But at the first sign of ripening, or even earlier, it should be wholly withheld."

The action of manure is even now very far from being thoroughly understood. When modern chemistry was first applied to agriculture, it was supposed that the great object of manure was merely to afford food for plants. But it was afterward found that other conditions were of equal importance, and that the advantage of many manures arose from their mechanical influence upon the soil. At Lois Weedon in England, excellent crops of wheat have been raised by thorough cultivation, without the application of manure, and the same principle was advocated by Jethro Tull in 1731, whose famous system of horse hoeing husbandry consisted simply in deep ploughing and thorough pulverization of the soil.

But while the mechanical condition of the soil exerts a most important influence on the growth of plants, there can be no doubt that unless all those elements of which a plant is composed, exist in the soil, or are derivable from other sources, healthy vegetation is impossible. Tull's farm finally failed to yield fair crops, notwithstanding large expenditures, on the mechanical part of the process, and the same result is said, to have attended the rigorous application of his principles elsewhere.

If the action of manures in general, is but imperfectly understood, still less does its influence upon the vine and its products, seem to have been reduced to known laws. In France the use of manure has been productive of evils so great as to induce the company of wine merchants, and vineyard proprietors, to condemn the use of azotized manures entirely.

On the other hand, the vine-dressers of Thomery, who produce the beautiful Chasselas de Fontainebleau grapes, use rich manures in liberal quantities. In general, it will be found in this, as in other cases, that a middle course is best. If the border has been purposely prepared in the first place, a vigorous growth will have been secured, while it was necessary that the vine should produce abundant wood, and when, after four or five years, the fruit is applied to the manufacture of wine, all rankness of growth will have disappeared. If, in after years, the vine should show symptoms of debility, it will be easy to supply it with nourishment, by means of liquid manure; and if ample means are provided for keeping the roots very dry during the ripening process, so that we can regulate the period over which the effects of such application shall extend, we are inclined to believe that no evil results will follow.

M. Ladrey suggests that but one portion of the vineyard be manured at one time, and that the wine from the part so treated be kept separate from the rest, until the evil influence of the manure has disappeared. It is obvious, however, that if we could avoid entirely any loss, of even a part of the vineyard, it would be desirable.

In this, however, as in all other matters, we must keep steadily in view the fact, that all rank vegetation exerts an injurious influence, not only upon the fruit product of the current year, but on the wood upon which our next year's crop depends.

In his "Nouveau système de la culture de la vigne," Persoz attempts to avoid the evils incident to the ordinary mode of the application of manure, by adding to the soil those matters which tend to produce wood, and those which favor the production of fruit, each at the appropriate time. His formulæ are as follows.

Six pounds bone dust; three pounds leather clippings and other animal refuse; (blood, horns, hoofs, etc.) and one pound gypsum, making in all ten pounds to be added to each square yard of border. This is done in the spring before the buds have pushed.

As soon as the young shoots are well advanced, he manures each square yard with eight pounds silicate of potassa, and two pounds of the double phosphate of potassa and lime. Silicate of potassa he procures by fusing fifteen parts of quartz sand with ten of potassa and two of charcoal.

The double phosphate of potassa and lime is prepared, by adding 18 lbs. of sulphuric acid to 24 lbs. of calcined and pulverized bones. This, after being well stirred, is diluted with water, allowed to stand for three days, treated with hot water and filtered. Carbonate of potassa is then added, until the liquid is slightly alkaline, and it is then evaporated in a cast iron vessel, roasted at a red heat, mixed with the silicate, and the whole reduced to powder.

A vine manured by Persoz with 0.5 kilogr. of silicate of potassa, 1.5 of phosphate of lime and potassa, and an equal weight of dried blood and goose dung, put forth in one year a shoot 11 metres in length, and yielded on nine shoots twenty-five bunches of grapes, while a similar vine, which was not manured, produced a shoot only 4.6 metres long, with only four or six blossoms, which faded away before their full development.—Liebig and Kopp: *Annual Report.*

CHAPTER XII.

DISEASES AND INSECTS.

We confess we have had very little experience in the matter of diseases and insects affecting the grape vine. Our native varieties are so vigorous and hardy that disease rarely affects them, and during the growing season they push with such rapidity, that the loss of a few leaves destroyed by insects is scarcely felt.

But we are aware that it is not always so, and we shall therefore give as full an account as we can obtain of the formidable pests to which the grape grower is exposed.

When growing in the open air in a suitable soil, and with a good exposure, the only two diseases to which the grape vine is liable, are mildew and the rot. The former appears in whitish spots on the surface of the leaves and wood, and when examined with a simple lens of 25 inch focus, shows a net-work of fungus with its sporules.

For this, as well as for the red spider, no remedy has been found equal to sulphur, the use of which for this purpose has been known from time immemorial.

To apply it, use may be made either of a common dredger fixed to the end of a pole, or of a pair of bellows with a contrivance for admitting a small quantity of sulphur into the stream of air. Or it may be mixed with water and the foliage syringed therewith. But the most efficient method is that proposed by Dr. Price, who was the first to suggest pentasulphide of calcium for this purpose.

This compound is prepared by boiling 30 parts by weight of caustic lime with 80 parts by weight of flowers of sulphur, suspended in a sufficient quantity of water; heat is applied until the solution has acquired a dark red color and the excess of sulphur ceases to dissolve. The clear solution is drawn off, and after being diluted with 20 times its bulk of water, may be applied to the vines by means of a sponge, brush or syringe.

Where flowers of sulphur is used, it should have a few drops of ammonia added before it is applied to the foliage, as the sulphurous acid with which it is saturated (derived from its combustion during distillation) is always injurious to leaves and young shoots.

The rot has rarely troubled our northern vineyards, though it is the great bane of vine culture in Ohio. We are inclined to believe that if vines are planted in soil, dry or well drained and not too rich, and be allowed to extend themselves moderately, but little need be apprehended from the rot.

Dr. Asa Fitch has found upward of thirty different insects which prey upon the grape vine, but with the exception of the red spider, and occasionally the rose-bug, they do not injure the vine materially.

The red spider (*acarus tellarius*) of which we give a cut (Fig. 60), we have found, not only under glass,

Fig. 60.

but on vines in the open air. It is a small, reddish-colored insect which it requires a sharp eye to detect. For this, as for mildew, sulphur is a specific, and we are always safe in giving our vines a good dusting of this substance, so as to *prevent* any injury which might arise from either source.

The rose-bug has never troubled us much. It nearly destroyed Dr. Underhill's vineyards at one time, however, and we therefore give his account of the matter in his own words.

"Several years since, when my vineyards were

smaller than at present, I found the rose-bug a formidable enemy. They appeared on the vines when they were in blossom, or just as the blossoms were falling off and the young grapes forming, and devoured them with the greatest avidity. This feast continued from eight to twelve days, or, until the cherries on the trees in the vicinity began to ripen, when they with one accord flew to them, for a change of diet, I presume, or from some other cause. I was quite familiar with the habits of the caterpillar, and had been in the practice of clearing them from my orchards in the spring, before they had destroyed scarcely a leaf. This I did not consider a great or difficult matter, for they were enveloped in a web early in the morning, and one man in a few days was able to clear many hundred trees, by twisting them off, web and all, with a basket, and carefully placing them under his foot. The rose-bug, however, did not, like the caterpillar, make its appearance in clusters or webs, but in small numbers at first, and scattered through the vineyards, increasing rapidly every day. Though taken from the vines on the trellis every morning, they continued to multiply till the eighth or twelfth day, when they suddenly left for the cherry-trees, as before stated. I was at a loss at first to know where they came from, till at length I discovered the ground perforated with numerous holes, through which they made their way to the surface.

"I observed, when they first appeared on the vines, they were so feeble as to be unable to fly even for a few yards. Having surmounted all other difficulties, I was determined not to be defeated in the vineyard cultivation of the grape by this insect, and consequently resorted to the following means for their destruction. I directed my men to take each a cup, with a little water in it, and go through the vineyards every morning, removing every bug from the vines; and this was done quite rapidly by passing the cup under the leaf, and merely touching it, when the bugs instantly dropped, and were received in the cup containing the water. When the cup was full, they were soon destroyed by pressing the foot upon them on a hard surface. After all of them had been taken off, on the following morning there were ten on the vines where we had found but one; and the succeeding morning, after having been removed as before, there were one hundred where there were but ten, and so on. I was not discouraged, however, and directed my men to persevere in the work of destruction, and we should thus perhaps prevent the formation of another progeny for the next season, for it is very easily shown that they do not migrate to any great distance; and by thus destroying the present race, I am convinced that we insure ourselves from their further depredations to any injurious extent. When a person of some energy has cleared them from his vineyard or garden, he is pretty certain to enjoy the benefit of his labor another season as well as the present, though he may have a few from his less resolute neighbor. Pursuing the course I have mentioned, I very soon lessened the rose-bugs so much that they gave me very little trouble.

"I also tried ploughing my vineyards just before winter set in, so as to expose to the weather the insect in the larva state, which will certainly destroy all the young tribe that have not descended below the reach of the plough. For two years past the number has been so small, that I have omitted this process for their destruction."

CHAPTER XIII.

METHOD OF HASTENING THE MATURITY OF THE GRAPE.

Several methods have been proposed for causing grapes to ripen at an earlier period of the season than usual, or in localities where they would not otherwise ripen at all. The most successful, and, on the large scale, economical, mode of effecting this is undoubtedly by means of glass houses, either with or without fire heat. A description of these is beyond the limits assigned to this work, though we may, perhaps, be allowed briefly to describe two devices of this nature, by which a few bunches may be matured at small expense and with very little trouble.

"More than twenty years ago, a market gardener at Bath published a plan of ripening grapes under common hand-glasses. He planted the vines in a soil composed in great part of lime rubbish; placed a glass over each plant, taking out half a pane in its summit through which the leading shoot of the vine protruded itself, and grew in the open air. The bunch or bunches of grapes remained within the hand-glass, and enjoyed the advantages of protection from cold winds, dews, and rains during the night, and of a high degree of confined solar heat during the day."—LOUDON.

Mr. Maund, editor of the "Botanic Garden," employs the following method of obtaining a few bunches: "Although my experiment is not yet completed, I cannot omit mentioning to you its success. Grapes grown on open walls in the midland counties are rarely well-ripened; therefore, I provided a small glazed frame—a sort of narrow hand-glass—of the shape shown in the annexed outline, to fix against the wall, and inclosed in it one branch of the vine with its fruit and foliage. The open part, which rests against the wall, is 13 inches wide, and may be of any length required to take in the fruit. The sides are formed of single panes of glass, seven inches wide, and meet on a bar which may represent the ridge of a roof, the ends inclosed by triangular boards, and having a notch to admit the branch. This was fixed on the branch a month before the vine came into flower. The consequence was, the protected branches flowered a week earlier than the exposed. The frame was not fitted closely to the wall, but in some places may have been a quarter of an inch from it. The lateral branches being shortened before it

Fig. 61.

was fixed, it did not require removal, even for pruning, because I adopt the long-rod mode of training, which is peculiarly adapted to my partial protection system.

"The temperature within the frame is always higher than that without, sometimes at mid-day even from 20 to 30 degrees. By this simple protection, I find grapes may be ripened from three weeks to a month earlier than when wholly exposed, and this saving of time will, I believe, not only secure their ripening well every year in the midland counties, but, also, that such advantage will be available in the north of England, where grapes never ripen on the open walls. I should have told you that the cold nights of spring have caused almost all the young fruit to fall off during the flowering season, excepting where it was protected.

"To hasten the maturity of grapes grown in the open air, means may be taken to throw them early into a state of rest. On the 20th of September prune the vine as you would in the month of December, taking off all the leaves and grapes, ripe or unripe, and shortening all the branches to one, two or three eyes at most. The following spring it will push its buds a few days before any of the neighboring vines pruned in winter. Train it as carefully all the summer as though you were certain it would ripen its crop of fruit. Pursue the same system annually, pruning the tree always between the 20th and 30th of September, and in the course of seven years you will be rewarded for your patience and expense with half a ripe crop in most summers, and a whole ripe crop in warm summers."—LOUDON.

The following method of hastening the maturity of grapes on open walls, was communicated to the Horticultural Society of London, by Mr. Thos. Fleetwood: "Before the vines are out of flower, he brings each bunch into a perpendicular position by a thread attached to its extremity, and fastened to a nail in the wall, carefully confining the young branch with the bunch thereon, as close to the wall as possible. The period of blossoming is preferred for this operation, because the bunch at that time takes a proper position, without injury.

By this practice the bunches are kept so steady that the berries are not bruised by the action of the wind, and being fixed close to the wall, they receive such additional heat, that they ripen a month earlier than when left to hang in the usual way."

But of all the plans which have been proposed, perhaps the simplest and most efficient is ringing, girdling or breaking. It has been employed for many years in France, although it is there conceded, that it injures the quality of the wine produced. For table purposes, however, the grapes seem to be improved both in size and appearance.

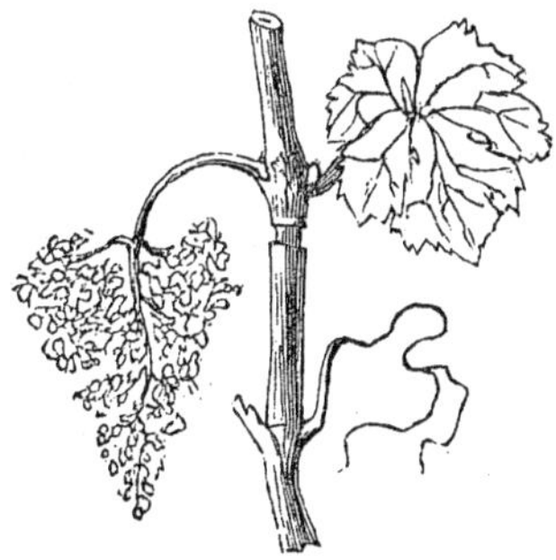

Fig. 62.

The French method is shown in Fig. 62. Here the annular incision is made just below the fruit bunch at the time of flowering. A pair of pincers with a double pair of semicircular jaws, makes both the upper and lower incision at once, when the bark is easily removed by the finger nail.

The following are the details of an English practitioner:

"The vines are generally cultivated upon the *Hoare* system, or, as it is called, the *long rod system;* but they are not so cultivated in every case, for sometimes an old bearer is spurred back to one or two buds, to carry its crop another year. My vines are very strong, and the rods or branches stand at least three feet, or even three feet six inches, distant from each other, when winter pruned. This allows just sufficient room for the fruit-bearing laterals and a young rod to come up between every two bearers. This young rod, of course, to be the bearer of laterals the following year:

"Thus no vines cultivated on any other system are so capable of being rung, without the disadvantage of killing or losing the future useful part of the tree, because on Hoare's long-rod system, the whole of the previous years, bearers will have to be cut entirely away.

"The very right time to perform the ringing is just after the berries are all set, or have attained the the size of No. 2 shot, or small peas. In ringing, cut, with a sharp knife, clean round the branch between two joints. Or, if you are going to ring the laterals carrying the fruit, leave either two or three buds and leaves beyond the main stem, and make the ring just in the middle, between the third and fourth leaves, or joints. As I said before, make two cuts clean through the bark, quite down into the wood, one inch apart, and remove the bark clean away, all round the branch or lateral. By this means, if you are in the habit of spur pruning, the hinder buds are left all right, to spur back the following year. If you prune upon the long-rod system, you may ring the rod just wherever you please— the whole branch, if you like— as the rung part will have to be cut away entirely after the fruit is gathered.

"The ringing is performed just the same on an old whole branch as in that of the young lateral carrying one or two bunches. I have repeatedly rung old branches, that have been carrying from twenty to thirty bunches of grapes, with the same good effect; only it has been such branches that I have intended to cut away entirely the following autumn: of course, thinning out the berries of the bunches, and the bunches too, if excellence be aimed at, is of the utmost importance. The process of thinning this cannot be too early attended to. I always begin as soon as the fruit is fairly set, and continue to remove all inferior berries, and, with a good pair of scissors and clean fingers, using my eyes to see what I am about, so as not to injure the berries by handling and mauling them.

"By thus practising ringing, I have produced for the last twelve or fourteen years, grapes, out of doors, that have puzzled many a tyro and others too.

"Our indefatigable editors have both watched my progress in vine culture for years. My grapes have many a time puzzled the late Mr. Elphinstone, when he was gardener to the late speaker of the House of Commons, now Lord Eversley, although I used to compete against him, with both indoor and outdoor grapes, at our Hampshire horticultural show in November.

"As a matter of course, I had read of ringing fruit trees, etc., but it never struck me to put the same into practice until about fourteen years ago, when my attention was called to it in an amateur friend's garden, Mr. Frampton, glass and paint merchant of this city. I happened to walk in and look at some vines to which he was paying great attention at that time. This was in the month of September, and here I first saw the ringing process of the vine. Seeing a few bunches of the *Black Hamburg* so large in the berry, and all ripe, I began to inquire into the particulars, when Mr. Frampton kindly showed me where the branches were rung, and that the ringing was the cause of their being so very large and so early. I then wanted to know whence Mr. Frampton obtained his information, when he showed it to me in the 'Penny Cyclopædia,' from the pen of Professor Henslow."—*Thos. Weaver, Gardener to the Warden of Winchester College.*

[It is quite true that we have watched for some years, with great interest, the experiment upon ringing vines carried on by Mr. Weaver, and we can authenticate his statement of the mode of ringing and its results. It must not be done in that petty timid manner hinted at by a contemporary. There must be a ring of bark perfectly removed; the cuts being made boldly down to the very young wood, or alburnum, and every particle of bark, inner and outer, must be removed between the cuts. (See Fig. 63.)

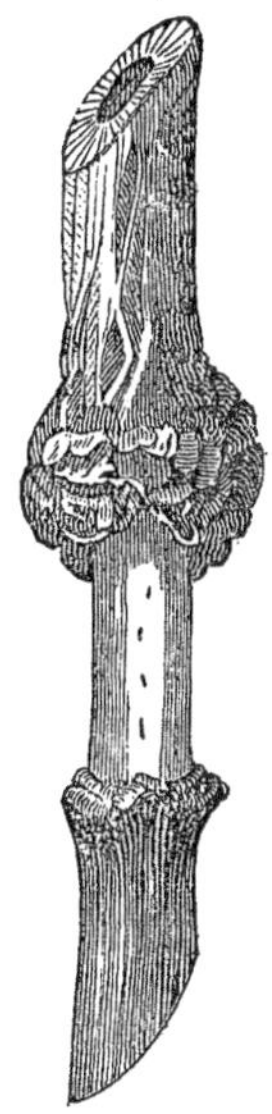

Fig. 63.

This drawing represents, faithfully, the rung part of a rod at the close of autumn, and shows how the removal of the band of bark checked the return of the sap, and how, in consequence, the rod above the removed band increased in size beyond that portion of the rod below the band.

The effect upon the berries was, in every instance, to advance their early ripening a fortnight, and to about double the size and weight of the berries, when compared with those grown on unrung branches of the same vine. Nor was the color and bloom of the berries diminished; indeed, so excellent were they, that we have seen them exhibited deservedly by the side of grapes grown under glass, and they were sold in November, at Winchester, for half-a-crown a pound.

Ringing the branches of fruit-trees, to render them fruitful, was practised in France, and recommended there in print, about a century and a half since. There are various letters upon the subject in the early volumes of the "Horticultural Society's Transactions," and in one of them (vol. 1, page 107), published in 1808, Mr. Williams, of Pitmaston, gives full directions for ringing the grape vine. He tells the result, in these words: "I invariably found that the fruit not only ripened earlier, but that the berries were considerably larger than usual, and more highly flavored."—*Editor of the Cottage Gardener.*]

CHAPTER XIV.

CARE OF OLD VINES.

THERE are scattered through the country numerous old vines of large growth and great age, which have been trained upon trellises, through trees, against the sides of houses and on arbors, without much skill or attention. These well deserve good culture, and the owners would gladly bestow it if they knew how. For their benefit, a few hints in this direction may not be out of place.

Such vines have in general either been left entirely to themselves, or trained wholly on the long-spur system, no new wood except these spurs being kept from last year's growth to supply the wants of next year, and the strength has thus been thrown to the ends of the stems, leaving them barren for a great distance from their base.

Vines in this condition, if of good origin, may, by judicious management, be speedily made to bear large crops of excellent fruit, as their roots are large and powerful, and fully competent to supply nutriment to a large crop of grapes.

If the stems are tolerably well supplied with bearing spurs, it may be advisable to take good care of such of these as we can find, and where there is a barren spot, to train a young shoot over it from the nearest bearing spur. Upon this young shoot spurs may soon be made, which will bear admirably.

But, in almost all cases, the better plan will be to gradually renew the whole vine, as strong, vigorous shoots, when once laid in for main branches and well supplied with bearing spurs or canes, will last for a long time and give satisfactory results with far less labor than is required by an old and straggling vine.

This change had better be effected gradually, a portion of the old wood being retained until the young shoots come into bearing, so that we need not be entirely deprived of fruit during its progress.

Commence, then, at the spring or winter pruning, and remove all the wood that can be well spared, keeping only a few of the best main branches, and cutting the spurs on these very close, leaving not more than one eye to each.

This severe pruning will cause the vine to throw up numerous strong shoots, or suckers, from near the roots. Two or three of the best of these must be selected and trained to stakes, away from the trellis or arbor, so as to give them all the light and air possible; the laterals which start from these must be pinched at the third leaf, and they should be stopped about the middle or end of September. All other shoots from the base of the vine, as well as all useless or barren shoots on other parts, must be carefully removed as fast as they appear, so as to throw as much as possible into the canes we had selected.

Next season, these canes must be disbudded and laid in as follows: Having removed all laterals and tendrils and tied them firmly to the trellis, as shown in Fig. 64, commence at the first good bud from the

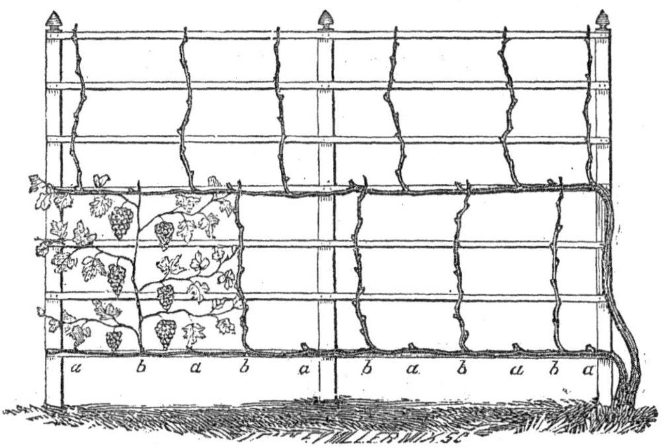

Fig. 64.

base, which leave, and then remove all the buds for a space of from 14 to 20 inches. Between 14 and 20 inches we will certainly find a good bud on the upper side of the cane (as it is tied to the trellis), which must be kept, and all the buds on the next equal space, removed in the same manner. So proceed until you have laid in ten or a dozen buds on each cane, when it should be cut off. We will now have two or more horizontal arms, each of which will throw up from 5 to 6 vertical canes of a strength sufficient to bear fruit next year, and the same number of short shoots which will form spurs for next year's bearing canes. But in order to make sure of this, we must prune the old vine very severely, indeed, and if we could make up our minds to do without fruit for one year and cut it *all* away, we would be gainers by it in the end. But in any case, all fruit must be removed from our new wood, as the stems will have enough to do to cover the trellis without bearing a crop of grapes.

Next year, the canes *b*, *b*, *b*, *b*, *b*, will bear a full crop of fruit, and shoots must be trained up from the spurs, *a*, *a*, *a*, *a*, *a*, *a*, to take their place at the winter pruning. The whole management will now be the same as that previously described for vineyards.

If it be preferred to train up the vine on the spur system, the buds at *a*, *a*, *a*, *a*, *a*, *a*, should be removed when the cane is disbudded the first season, and after having borne once on the long-rod system, the canes *b*, *b*, *b*, *b*, *b*, will be well provided with shoots by cutting back on which good spurs may be formed. These spurs should be distributed along the canes at a distance of 14 to 20 inches on each side, and may be managed individually, as described in Chap. VII.

The height to which spur-bearing canes may be carried is, perhaps, without limit, if they are properly treated and the vines have sufficient root power. But in practice, we do not think that it will be well to have them longer than 6 to 8 feet. They are thus kept within bounds, and any one which may become barren is more easily renewed than if they are of greater length.

Where the vines are managed on the long-rod system, we would never have the canes over 6 feet long, and if only 4½ to 5 feet, so much the better.

Thus, if we desired to cover a wall or trellis fifteen feet high, we would have two tiers of arms carrying spur-bearing canes each 7 feet long, or three tiers carrying long-rod or renewal canes.

Before proceeding to renew an old vine, it may be well to manure it thoroughly, either by a good top dressing in the fall, liquid manure during the growing season, or by digging a trench about six feet from the roots and filling it with good compost, bones, etc.

An excellent plan for feeding an old vine is to make a basin about six inches deep round its roots, with boards, against the outside of which sufficient heavy soil has been placed to make it water-tight. Then, during the growing season, let this basin be filled with soap-suds every washing day—mixing them with chamber slops, etc. During the winter, it should be filled with leaves and prunings, over which a little earth may be thrown to keep the wind from blowing them about, and preventing an unsightly appearance. If the roots of the vine are so near the house as to be unsightly when treated in this manner, the basin might easily be provided with a light board cover neatly painted. It might be requisite to form it in two parts, having notches through which the stem of the vine can pass.

That a good manuring will often cause a vine which has been previously unfruitful to bear abundant crops, is well known. We have now in mind an instance of a vine which after remaining barren for many years suddenly became quite fruitful from chickens making a roost of the trellis on which it grew.

CHAPTER XV.

TO PRESERVE GRAPES.

ALTHOUGH He who "has made everything beautiful in his season," no doubt designed grapes to be used while fresh, yet, though we cannot preserve the exquisite flavor of newly-gathered grapes, we may, nevertheless, prolong their season, if not in its full excellence yet with sufficient attraction to make it worth while.

With proper care, grapes may be kept until Christmas, and at that time will command a price which would not be paid for fresh fruit during the height of the grape season.

As yet, the preserving of the fruit seems to be but little understood, and although we have kept grapes until January in a very palatable state, and we have tasted others which have been tolerably preserved until March, we must acknowledge that none of these attempts quite came up to our desires, however much others might have praised the result. The truth is, that grapes in March will never be very severely criticised under any circumstances. They are too much of a rarity for that.

Although the foreign grapes which are imported, packed in sawdust, are said to be gathered before they are fully ripe, we believe that this plan is not suited to our native varieties. They should always be fully ripe before they are gathered, and this should be done on a clear, dry day before they have been touched with frost. The bunches should be carefully examined, none but the first-rate ones selected, and they must be scrupulously freed from all dirt, such as leaves, spiders' webs, insects, etc. All decayed or unripe berries must be removed with a pair of sharp scissors (merely pulling them off will not do); and they should be exposed to the air (but not the sun) for a few hours before being packed away. In one case where, after the grapes were gathered, the weather became damp before they were put up, we know them to have been placed in a *moderately* warm oven for rather more than five minutes, and the result was very good.

The following are a few of the methods which have been recommended:

1st. Procure some fine, dry sawdust (avoiding that from resinous or scented wood), and pack the grapes in a box or barrel, in layers, being careful to have sufficient between the bunches to prevent their touching. Bran is sometimes substituted for sawdust.

2d. Wrap each bunch in fine, clean dry paper, and put away in layers in boxes.

3d. Take a good box and place a layer of cotton batting on the bottom; on this place a layer of grapes, then a layer of batting and so on, until the box is full, wrapping each cluster in thin paper. Some omit the paper.

4th. Seal up the ends of the stems with wax, and suspend them in a cool, dry and dark room, looking them over occasionally and removing unsound berries and bunches.

The French suspend their bunches by the lower ends to a little hook (see Appendix). Some cultivators, however, cut away the fruit-bearing branches and preserve the grapes attached to them.

It has been advised to immerse the stems of the bunches in wine, before the fruit is used; but as they are always dried up and incapable of transmitting fluid, we have found it better to immerse the whole bunch in cold water for half an hour or so. This restores the plumpness of the berries and removes some of the foxy flavor which is apt to tinge our native grapes when long kept.

CHAPTER XVI.

DESCRIPTIVE LIST OF NATIVE GRAPES.

BLAND.

Synonyms—Bland's Madeira, Bland's Pale Red, Bland's Fox, Bland's Virginia, Carolina Powel, Red Bland, Red Scuppernong. Where this grape will ripen well it is valuable on account of its fine flavor. It is, however, confined to the most southern and favorable localities, of which we take cognizance in this work, rarely ripening north of the Hudson, although Elliot states that, in 1820, it was well grown and ripened in New Haven, Connecticut, in sandy soil.

Said to have originated in Virginia, discovered by Col. Bland, who presented cuttings to W. Bartrem, the botanist, and also to Samuel Powel, Esq., after whom it was in some cases named.

Bunches shouldered, long, loose. Berries round or slightly oblate, medium size, pale red when ripe, juicy, sweet sprightly flavor, very little pulp. Foliage pale green, smooth and delicate.

CANADIAN CHIEF.

This grape is claimed to be a native of New Jersey; but our best pomologists are of opinion, that if a native, it is at best but a seedling, from some foreign variety.

The bunches are large, berries medium, green or faint amber, and the flesh tender.

In Canada it is said to be perfectly hardy and to ripen well in the open air. Grape-growers in the United States have sometimes found that it is apt to be winter-killed, and that it mildews badly.

CATAWBA.

This is the great wine grape of the South. It was first introduced by Major Adlum, of Georgetown, D. C., and has been subsequently patronized by N. Longworth, Esq., the father of American wine culture.

Bunches medium size, loose, shouldered. Berries large and round or very slightly oval. Skin rather thick, pale red in the shade, deeper red in the sun, and covered with a lilac bloom. Juicy, sweet, musky. Should be allowed to hang till fully ripe. Downing states, that *unless* fully ripe it is more musky than the Isabella. Prince, on the other hand, says, that when fully ripe it is quite musky. Our own experience leads us to think that it is more musky when ripe than when unripe.

CLINTON.

It is generally believed that this grape originated in western New York. It is extremely hardy and productive, but as a table fruit we regard it as scarcely worthy of cultivation. It is said to ripen several days before the Isabella, but until well touched with frost it is uneatable. As a wine grape, however, it is said to be unequalled, amongst those grapes which ripen, where the Catawba fails to come to maturity.

Bunches medium or rather small, shouldered, compact. Berries small, round, black, thick bloom, juicy, acid and astringent.

CONCORD.

Though by no means a fine grape, the Concord is valuable from its quality of ripening ten days or so before the Isabella, and consequently maturing in a large range of country where that grape fails. It is very vigorous, hardy, and productive. C. Downing describes it as follows:

"Bunch rather compact, large, shouldered. Berries large, globular, almost black, thickly covered with bloom. Skin rather thick, with more of the native pungency and aroma than the Isabella, which it resembles, but does not quite equal in quality. Flesh moderately juicy, rather buttery, very sweet, with considerable toughness and acidity in its pulp."

The Concord grape becomes more foxy the longer it is kept, hence two persons, one of whom ate the fruit fresh from the vine, and the other obtained it only after it had been gathered some time, might form very different ideas as to its quality.

DELAWARE.

This fine grape, which promises to stand in the front rank of our hardy native grapes, is said to have originated in New Jersey, whence it was carried to Ohio, and falling into the hands of A. Thompson, has been thence distributed pretty widely amongst fruit-growers. Some German vine-dressers have supposed it to be the Traminer, while others have thought it the Resling, but we believe our best pomologists are agreed that it is a native. The following description is by C. Downing:

"Bunch small, very compact, and generally shouldered. Berries smallish, round when not compressed. Skin thin, of a beautiful bright red or flesh color, very translucent, passing to wine color by long keeping. It is without hardness or acidity in its pulp, exceedingly sweet but sprightly, vinous, and aromatic, and is well characterized by Mr. Prince, as our highest flavored and most delicious hardy grape."

DIANA.

A seedling of the Catawba raised by Mrs. Diana

Crehore, of Boston, and named after her by the Massachusetts Horticultural Society. Next to the Isabella and Catawba, the merits of this vine are perhaps the best established of any we have. It is a vigorous grower, a productive bearer, and extremely hardy. The bunches are large, the berries but slightly less than the Catawba, and of about the same color, perhaps a shade darker when ripe. The flavor is very superior, and even before being fully ripe is still quite good, and is esteemed by some, as even then superior to the Isabella. Another excellent quality consists in the ease with which the fruit may be kept for winter use.

ELSINBOROUGH.

A fine though small table grape, found near Elsinborough, Salem Co., New Jersey, and first introduced by Dr. Hulings. It is hardy and productive, and worthy of more general cultivation than it has received. Bunches medium, loose, shouldered. Berries small, round, black and covered with a blue bloom. They have generally but two seeds and are free from pulp or musky taste.

Different authors have different modes of spelling the name of this grape. Thus it is called Elsinburg, Elsinburgh, etc. We prefer to spell it in the same manner as the village from which it takes its name

KING.

This is a new grape which sprung up accidentally in a garden in the northeastern part of this city, and is supposed to be a seedling of the Clinton, which it resembles very much both in habit and foliage; the fruit, however, being of a green or yellowish hue instead of black. We first saw it in bearing, in the fall of 1857 and thought so much of it as to procure cuttings. The fruit was brought before the Fruit-growers Society in 1858, and named by them in honor of Wm. King, by whom it was introduced. It appears to be as hardy as the Clinton, and as the flavor is much superior it cannot fail to prove a valuable acquisition.

HARTFORD PROLIFIC.

A hardy, vigorous and productive variety which originated in Connecticut, and matures in latitudes where the Isabella and Catawba fail to ripen. Bunch large, shouldered, compact. Berry large, round, with thick black skin covered with bloom. Sweet, juicy and acid, but with a good deal of the native perfume. Ripens ten days before the Isabella. The berries have sometimes been found to fall from the bunch as soon as ripe, leaving a number of unripe berries. Lately, however, this difficulty has been lessened by superior cultivation.

HERBEMONT.

This is one of the grapes recommended by the American Pomological Society, as promising well. There is so much discrepancy in the descriptions by various authors, that we confess some doubt as to the identity of the varieties described. We quote the following from C. Downing: "This is the most rampant grower of all our hardy grapes, and under favorable circumstances yields a fruit of surpassing excellence, with which the nicest detector of foxiness, thickness of skin, toughness or acidity of pulp, can find no fault; north of Philadelphia, it needs a warm exposure or favorable season for the full development of its excellences. In our village, under the care of a lady, it has not failed for many years to give a most abundant crop of perfectly ripened fruit, and without protection has not suffered at all from winter-killing. A very old vine in Baltimore, which had never before failed to produce abundantly since its first bearing, had, last winter, when the mercury fell to 19° below zero, all its young wood killed; but ordinarily in that latitude and further south, it is an unfailing bearer and particularly fitted for those southern latitudes that are liable to injury from late frosts in spring and early frosts in autumn, as it flowers very late and ripens its fruit early. Its leaves in autumn are the last to yield to frost, remaining perfectly green and vigorous after all others have withered and fallen; consequently, it has often an amount of unripened wood, which should be cut off before winter.

"Bunch very large and exceedingly compact, shouldered. Berries below medium, round, dark blue, or violet, covered with a thick, light bloom. Skin thin, which is filled with a sweet, rich, vinous aromatic juice, of so little consistence that it cannot be called flesh.

"LENOIR, LONG, DEVEREAUX, AND THURMOND.—Under the above names, grapes much resembling in character the Herbemont, are grown in the southern States, and we have hitherto considered them synonymous of it; but our southern friends claim that Lenoir is a distinct variety, and much earlier than any of the others, and also that at least some of the others are distinct. The matter is now under investigation, and we must wait the result before deciding."

ISABELLA.

The popular account of the origin of this vine, is that it was a native of South Carolina, which being brought to the North and introduced to the notice of cultivators by Mrs. Isabella Gibbs, the wife of George Gibbs, Esq., was named in honor of that lady. It has, however, been attempted to throw some doubt upon this history—not, perhaps, as to the facts themselves, but as to their accounting for the origin of the Isabella grape—some pomologists claiming that it is a widely distributed and well-known native species, while others assert that it is a well-known European variety.

But be this as it may, it is certainly one of our most prolific and vigorous varieties, and is thus far more widely cultivated at the North than any other with which we are acquainted. Nor is this preference misplaced, as it excels all others which have been fully tested, both in the amount and in the certainty of the crops produced. When quite ripe, the flavor also is excellent, and the pulp almost disappears. Few realize this condition, however, as the fruit is generally gathered long before it is ripe. The following description is by A. J. Downing:

"Bunches of good size, five to seven inches long, rather loose, shouldered. Berries oval, pretty large. Skin thick, dark purple, becoming at last nearly black, covered with a blue bloom. Flesh tender, with some pulp, which nearly dissolves when fully mature; juicy, sweet, and rich, with slight musky aroma."

LOGAN.

This is a new grape which is recommended by the American Pomological Society as *promising well.* It ripens about the first of September; is black, sweet and good. Bunches and berries large.

A. Thomson, Esq., of Delaware, Ohio, describes it as follows: "It is a black grape, ripening before the Catawba, and preferred to the Isabella, and is believed to be a wilding of Ohio; hardy, vigorous; wood short-jointed and compact; distinct in wood and foliage, productive, and probably the earliest hardy grape of fair quality in cultivation, and will ripen its fruit several degrees further north than the Isabella and Catawba."

EARLY NORTHERN MUSCADINE.

With regard to this grape, the most contradictory accounts have been published. Some pomologists speak of it in high terms, and by others it has been as fiercely condemned. A gentleman, in whom we have every confidence, assures us that wherever he has introduced it it has given satisfaction, and we believe that he has no peculiar interest in this particular variety. Our own experience is not sufficient to warrant us in giving a decision, and we therefore append a statement by Messrs Lewis and Brainard, agents for the Shaker Society. We have, however, tasted wine made from this grape, which gave promise of much excellence. We have under cultivation a vine procured direct from New Lebanon, and expect ere long to satisfy ourselves in regard to its merits.

"THE EARLY NORTHERN MUSCADINE,

"*A Shaker Seedling,*

Of which the accompanying plate is a fac-simile of the cluster, ripens 15th of September; light amber color, medium size; delicious flavor, many say unsurpassed.

"This excellent grape, the subscribers affirm, ripens nearly a month earlier than the Isabella in the same latitude; is perfectly hardy for the northern climate; a sure and constant bearer; if properly and judiciously pruned, bears enormously; not subject to mildew, slightly fibrous in pulp, and has often been pronounced, by competent judges, superior in its season, to the Isabella or Catawba in their season, either as a table or wine grape. Its characteristic is peculiarly that of a summer fruit; and wine made from it, simply with the addition of sugar, has been often pronounced by hundreds superior to the best Sicilian light wines, which it somewhat resembles;

and very high prices have frequently been offered and refused for it, by those who were acquainted with its merits from actual use.

"Prof. J. P. Kirtland, M.D., of Cleveland Medical College, a correspondent of the 'Ohio Farmer,' of Cleveland, Ohio, Nov. 7, 1857, made the following remarks: 'During the last three weeks, we have amused ourselves in treating, perhaps, a hundred individuals to specimens of the Northern Muscadine, Catawba, Diana, Clinton, Isabella and Winslow seedling. Four in five (or four-fifths) of these persons have decided the Northern Muscadine to be the best in that list.'

"The subscribers affirm that twenty-five years' trial of this grape, in connection with about forty other kinds of our best modern, foreign and domestic grapes, give the Muscadine a large superior margin of profit. In short, its merits only need to be known to be appreciated, however much it may have been demerited by pomologists entirely unacquainted with its quality. It has taken premiums in several fairs in the United States, and has never, in our knowledge, been condemned by those who have raised it and tasted it fresh from the vines, or when properly kept, though it is not a long-keeping variety; but in this respect, is like all our choicest summer fruits.

"The best recommendation for this grape is, that

all who have ever raised the genuine Northern Muscadine, speak well of it, while it is constantly sought after where best known, in preference to all other varieties, notwithstanding some pomologists, unacquainted with it, have decided against it for reasons best known to others than to the suscribers.

"Multitudes of spurious varieties have been sold for this grape.

"JESSE LEWIS, &
"D. C. BRAINARD. } Agts."

REBECCA.

Described by the Committee on Native Fruits, of the American Pomological Society, as follows:

"The Rebecca originated in the garden of Mr. E. M. Peake, Hudson, N. Y., about eight years ago. Mr. Peake's garden is in one of the thickly settled streets of the city, and nearly the usual size of a lot, perhaps one hundred by one hundred and fifty feet deep. Between the house and the street there is a small flower garden. It was here that the original vine grew. Mrs. Peake was about making some alterations in her flower-beds, and this vine being in the way, her gardener advised her to dig it up, as it was only an 'old wild grape.' But disliking to destroy it, she removed it with her own hands, and planted it very carelessly in the garden, back of the house, in a very poor and cold clay soil. Here the vine made slow progress, but continued to increase in size until the third or fourth year, when it produced a few clusters of small white grapes. These appeared to possess so much merit, and were so much better than had been expected, that pains were then taken to feed and nourish it, and prune it into shape, and it soon well repaid all the labor bestowed upon it. It grew vigorously, making shoots ten or fifteen feet long, and bore abundant crops of the most delicious grapes, until at the present time it has reached the top of the house, and covers a trellis ten feet wide and twenty-five feet high, loaded with fruit.

"Bunches medium size, about six inches long, very compact, without shoulders; berries medium size, obovate; about three quarters of an inch in diameter; skin thin, greenish white, becoming of a pale amber color at full maturity, covered with a thin white bloom; flesh very juicy, soft and melting, and free from pulp; flavor rich, sugary, vinous, and brisk, with a peculiarly musky and luscious aroma, distinct from any other grape; seeds small, two to four in each berry; leaves scarcely of a medium size, about seven inches long and seven in width, very deeply lobed and coarsely and sharply serrated; upper surface light green, slightly rough; under surface covered with a thin, whitish down; nerves prominent; petioles rather slender." (See report of Sixth Session of Am. Pom. Soc.)

TO KALON.

Some pomologists have characterized this variety as a very poor bearer, while by others it is highly recommended. At the sixth session of the Pomological Society it elicited considerable discussion.

C. Downing speaks of it as follows: "This fine grape has been but little disseminated in consequence of the general supposition, that is was very much like, if not identical with the Catawba, from which it is entirely distinct in wood, foliage and every characteristic of the fruit. It is a vigorous grower, foliage very large, abundant, and much less rough than Catawba, or Isabella, and the alæ of the leaves overlap each other differently from any other with which we are acquainted.

"Bunches large and shouldered. Berries varying in form from oval to oblate, very dark in color and profusely covered with bloom. Its fruit when ripe is sweet, buttery and luscious, without foxiness in its aroma, or any toughness or acidity in its pulp. It is perfectly hardy, and with good treatment in deep, rich, pervious soil, it is an early and abundant bearer; with indifferent treatment it is a poor bearer. It ripens a little earlier than the Isabella."

It is one of the seven varieties recommended by C. Downing for general cultivation, but it is not recommended as even promising well by the Pomological Society.

UNION VILLAGE.

A fine black grape, said to resemble the black Hamburgh very much. It is a vigorous grower, hardy and productive. A little earlier than the Catawba or Isabella. Recommended as promising well by the Pomological Society.

NEW VARIETIES.

New varieties of grapes may be raised by the methods formerly described. Most of those so produced will of course prove worthless, being seldom equal to the parents. But occasionally a fine variety will reward our efforts and afford ample compensation for a thousand failures. Amongst those who devote their attention to the raising of new kinds may be mentioned, J. Fiske Allen, Esq., of Salem, Massachusetts, and Dr. Valk, of Flushing, Long Island. O. T. Hobbs, Esq., of Randolph, Pa., also informs us that he intends to devote his entire nursery to the production of new kinds of fruits and flowers, making the hardy native grape a specialty. He has already produced two new varieties which are said to be of considerable excellence, viz. the Kitchen grape and the North America, both seedlings from the Franklin grape, which is also a variety first brought forward by Mr. Hobbs.

LISTS OF SELECT GRAPES.

BY THE AMERICAN POMOLOGICAL SOCIETY.—*For general cultivation.*—Catawba, Concord, Delaware, Diana, Isabella. *Grapes which promise well.*—Herbemont, Logan, Rebecca, Union Village.

LIST BY C. DOWNING.—Catawba, Isabella, Diana, Delaware, Rebecca, To Kalon, Concord.

Our own views are that the Isabella is the most reliable grape for general cultivation at the North. By proper culture, both the Isabella and Catawba may be grown much further north than they are usually found, and it is possible that amongst our new varieties some may be found to excel there. The Diana is now pretty well established, and is a most excellent variety, and we would by all means encourage a trial of the Delaware, which gives great promise. For a garden, we should choose Isabella, Diana, and Delaware with the Catawba when it will ripen. It would also be well to plant a Clinton where the exposure is not sufficiently good to warrant the planting of a better variety. The Clinton is a hardy grape which will mature where other kinds will not, and although the fruit is not eatable in our estimation, yet it makes good wine. The King grape, however, seems to be quite as hardy as the Clinton and the quality promises to be equal to many of the more celebrated kinds.

Where the vines we have named will not succeed, we would plant Concord, Northern Muscadine and Hartford Prolific in preference to any other kinds, and under any circumstances we confess to a penchant for variety, as it is often found that the good and bad qualities of any particular grape, are confined to certain localities, so that by extending our selection we may gain an experience which will enable us to avoid the evil and to obtain the good.

TABULAR DESCRIPTIVE CATALOGUE

OF

NATIVE GRAPES.

The following catalogue makes no pretensions to originality, the descriptions given being in general those published by the originators of the variety, where such descriptions were procurable. The numerous blanks occurring in the tables will show at a glance the extreme imperfection of the descriptions usually published. Instead of giving such an account as would enable us to recognize the variety or to appreciate its real qualities, most so-called descriptions consist merely of a string of eulogies. The reader will observe that sometimes (for want of more definite information), we have described the flavor as "good," "pleasant," etc., although such a description is no description at all, not only because tastes differ so much as to what is good, but because there probably never was a grape which was not thought by its originators to excel everything else. We give their statements merely for what they are worth.

In many cases we have been able to give the name and nothing more. If by so doing, however, we can lessen in a slight degree the further increase of synonyms, the space thus occupied will not be wasted. There are certain names which seem to occur to every one who has or thinks he has a new variety and the bantling straightway receives a name which has probably served half a dozen before it.

CONTRACTIONS.—(bk.) Black; (s.) Shouldered; (d. b. C.) Days before Catawba; (I.) Isabella.

NAME.	SYNONYM FOR.	BUNCH.		BERRY.		COLOR.	FLAVOR.	RIPENS.	REMARKS.
		SIZE.	SHAPE.	SIZE.	SHAPE.				
Anna............	....	large.	shouldered	large.	globular.	white.	sweet, rich.	with Diana.	Vigorous and hardy
Albino..........	....	medium.	....	medium.	....	white.	good.		
Albiness.........	Albino.								
Alexander's	....	....	compact.	medium.	oval.	black.	sweet, musky	last of Oct.	
Alexander's York Madeira	Alexander's, Canby's Aug of some.								
Ada..	....	large.	compact.	....	....	black.	sweet, vinous		
Amanda.........	....	small.	....	medium.	....	purplish bk.	sweet, foxy.	1st Sept.	Hardy.
Amber Catawba..	....	medium.	....	medium.	....	amber.	sweet, musky	15 d. b. C.	
Amer'n Hamburgh	....	large.	....	large.	....	black.	foxy.	....	Hardy and vigorous
Amer'n Muscadine	Scuppernong								
Arkansas	Catawba.								
August Coral.....	....	....	....	large.	round.	bright red.	sweet.	end of Aug.	Hardy.
Baldwin	Canby's Aug								
Baldwin's Early..	Canby's Aug								
Baxter	....	very large.	long, com.	medium.	....	black.	acid.		
Bartram.........	....	large.	compact.	large.	round.	dark purple.	sweet, foxy.	....	Hardy and prolific.
Belle Isle	....	....	....	medium.	round.	pale red.	foxy.	....	Leaf variegated.
Black Claret.....									
Black Guignard..	....	....	oblong, s.	small.	round.	black.	sweet, vinous		
Black King	Summer Bk?	medium.	shouldered	small.	....	black.	foxy.	last of Aug.	
Black Spanish....	Montgomery								
Blackstone.......	....	....	....	large.	round.	black.	coarse, foxy.	middle Sept.	
Bland	....	medium.	loose.	large.	round.	pale red.	sweet.		
Bland's Madeira..	Bland.								
Bland's Pale Red.	Bland.								
Bland's Virginia..	Bland.								

Blood's Black				large.		black.	foxy, good.		
Blood's White				large.		white.	foxy, good.		
Brinckle		large.	compact, s sometimes.	large.	round.	black.	sweet, vinous		
Brown Guignard						chocolate.			Not eatable, but said to be good for wine.
Bull	Scuppernong								
Bullet	Scuppernong								
Bullitt		medium.	compact.	medium.		white.	sweet.		
Burton's Early		small.		large.	round.	black.	foxy.	middle Aug.	
Bula Scuppernong									Resembles Scuppernong, but larger & less palatable.
Bullace	Scuppernong								
Bullett	Scuppernong								
Camak				small.	round.	brownish red	sweet.		
Canby's August		medium.	compact.	medium.	roundish.	black.	sweet.	8 d. b. I.	
Cape / Cape Grape	Alexander's, Canby's Aug of some.								
Carolina Bk Cluster				small.	round.	black.	vinous.	end of Sept.	
Carolina Blue Clus'r				small.	round.	blue.	vinous, acid.	middle Sept.	
Carolina Blue Musk				small.	ovate.	blue.	vinous, acid.	middle Sept.	Very musky.
Carolina Globose			loose.	medium.	round.	black.	v., very acid.	end of Sept.	
Carter	To Kalon.								
Carter's Favorite			shouldered	medium.	round.	black.	sweet.		
Carter's Virginian				small.	ovate.	purple.		middle Sept.	
Cassady		medium.	compact.	medium.	round.	greenish white.	juicy, pleasant.		
Catawba		medium.	shouldered	large.	round.	pale red.	sweet, foxy.		
Catawba Tokay	Catawba.								
Catherine									
Cecilia		medium.		medium.	round.	black.	foxy.	middle Sept.	
Charleston									
Charter Oak		small.		large.	round.		foxy.		
Chippewa									Said to be barren.

CONTRACTIONS.—(v.) Vinous; (a.) Acid; (d. b. I.) Days before Isabella.

NAME.	SYNONYM FOR.	BUNCH.		BERRY.		COLOR.	FLAVOR.	RIPENS.	REMARKS.
		SIZE.	SHAPE.	SIZE.	SHAPE.				
Chocolate........	....	large.	....	small.	....	black.	acid.		
Christea's Isabella									
Chillicothe Seedling	....	....	long, loose.	medium.	oval.	dark purple.	....	middle Sept.	
Clara.............	....	medium.	loose.	medium.	round.	green.	rich & sweet	....	Salmon in sun.
Clarence	Lenoir.								
Clarke...........	....	....	....	large.	....	black.	acid.		
Clermont.........									
Clifton Constantia	Alexander's, Canby's Aug of some.								
Clinton	....	medium.	compact.	small.	round.	black.	acid.	10 d. b. I.	Very hardy; good for wine.
Cobstovine.......	....	....	....	....	....	....	....	....	Resembles Isabella.
Collina..........	....	....	compact.	small.	....	bluish black.	a., delicious		
Columbia	....	small.	compact.	small.	....	black.	pleasant, v.		
Columbian Muscat	....	....	....	large.	....	pale green.	high musk.	end of Sept.	
Concord	....	large.	shouldered	large.	round.	black.	sweet, foxy.	14 d. b. I.	Vigorous and hardy
Constantia.......									
Corbin	....	....	....	medium.	....	....	sweet.	....	Hardy.
Coriel	....	medium.	....	small.	oval.	bluish black.	good.	....	Hardy.
Creveling	....	....	....	small.	round.	black.	sweet.	14 d. b. I.	
Crystal	....	medium.	compact.	....	round.	white.	sweet.	....	Vigorous.
Cunningham.......	....	small.	....	small.	round.	black.	good.	middle Oct.	Hardy and vigorous
Delaware	....	medium.	compact.	small.	round.	pale red.	sweet, vinous		
Del'ware Burgundy									
Devereux.........	....	medium.	compact.	small.	....	purple.			
Diana	....	medium.	shouldered	large.	round.	reddish lilac.	sweet, vinous	10 d. b. I.	

Early Amber.....	....	....	....	medium.	oval.	amber.	sweet, foxy.	....	Hardy.
Early Black Fox..	....	....	....	large.	round.	black.	sweet, foxy.	1st Sept.	Hardy.
Early Red Fox...	....	....	....	large.	round.	red.	sweet, foxy.	10th Sept.	Hardy.
Early White Fox..	....	....	....	large.	round.	white.	sweet, foxy.	10th Sept.	
Early Isabella....	....	....	....	....	....	....	....	....	Resembles Isabella, but said to be earlier. This is probably owing to accidental circumstances.
Early Northern Muscadine	Northern Muscadine.								
Early Hudson....	....	large.	....	large.					
Early York									
Elizabeth	....	....	....	medium.	round.	black.	....	middle Sept.	Seedling of Isabella
Elizabeth	Hart's White								
Eaton's Catawba.	....	....	....	large.	round.	red.	sweet.		
Elkton...........	....	....	....	large.	....	purple.	foxy.		
Ellen	....	....	....	large.	....	dark purple.	sweet.	same as I.	
Elsinbro	Elsinburgh.								
Elsinburg.........	Elsinburgh.								
Elsinburgh.......	....	medium.	loose.	small.	....	black.	sweet.	....	Without pulp, h'rdy
Emily	....	large.	loose.	small.	round.	pale red.	sweet.		
Empire...........	....	large.	compact, s	large.	....	dark.	sweet, rose fl.	....	Hardy and vigorous
Estelle.........	....	....	....	large.	round.	black.	acid, foxy.	1st Sept.	Hardy and vigorous
Eudora	....	....	....	....	....	....	foxy.		
Eugenia	....	....	....	medium.	round.	dark red.	....	12 d. b. C.	
Eureka	....	medium.	not s.	medium.	round.	pale pink.	sweet, juicy.		
Fenner's Early...	Canly's Aug.								
Fox Grape (of the South)	Scuppernong								
Fragrant Harrison	....	large.	....	large.	....	amber white.	....	before I.	Seedling of Isabella
Franklin.........	....	....	....	....	....	....	....	....	Resembles Clinton, but better flavor.
Garver's Seedling..	Mary Ann.								
Garrigues	....	large.	shouldered	large.	oval.	dark purple.	sweet, rich.	8 d. b. I.	

CONTRACTIONS.—(v.) Vinous; (a.) Acid; (s.) shouldered; (d. b. I.) Days before Isabella.

NAME.	SYNONYM FOR.	BUNCH.		BERRY.		COLOR.	FLAVOR.	RIPENS.	REMARKS.
		SIZE.	SHAPE.	SIZE.	SHAPE.				
Gibbs Grape	Isabella.								
Golden Clinton	King.								
Graham		medium.	shouldered	small.	round.	purple.	saccharine.		
Godman		medium.	shouldered	medium.	round.	black.	sweet.		
Gridley		large.	shouldered	small.	round.	blue.	pulpy.		
Halifax				medium.	round.	black.	juicy, vinous		Unpleasant, but makes good wine.
Halifax Seedling				medium.	round.	black.	pulpy.		Flavor resembles Huckleberry; makes good wine.
Hartford Prolific		medium.	shouldered	medium.	round.	black.	sweet, foxy.	10 d. b. I.	
Hall's Grape				medium.		dark.		early.	Resembles Clinton.
Hannah			compact s.		oval.	dark purple.	sweet, musky	same as I.	
Heath	Delaware.								
Harris				medium.	round.	black.	sweet, pulpy	1st Sept.	
Herbemont		large.	compact.	small.	round.	dark blue.	sweet, vinous	early.	
Herb'mont Madeira	Herbemont.								
Hickman Grape	Scuppernong								
Hill Grape	Collina.								
Holmes		medium.	shouldered	large.	round.	purple.	sweet, foxy.	Aug. 25.	
Honey	Raabe.								
Hudson				large.	round.	black.	sweet.		Seedling of Isabella which it resembles
Hunterville			s. loose.	small.	round.	black.	acid, juicy.	end of July.	Good for wine.
Husman's Prolific			long, com't	small.	round.	black.	pulpy, astri't	end of Sept.	Good wine grape.
Hyde's Eliza									Said to be the same as Isabella but ripens 10 d's earlier

Illinois Prolific		medium.	compact.	very small.		juice very dark.			Makes a good red wine.
Imitation Hamb'gh	Union Vill'ge								
Isabella		medium.	shouldered	large.	oval.	black.	sweet, rich, foxy.		
Ive's Seedling				small.			foxy.		
Jack Grape	Ohio.								
James' Seedling									
Jennings' Seedling		medium.				black.		very early.	Ordinary quality.
Kauffman			compact.				acid.		Resembles Clinton.
Keller									
Kendall		medium.		large.	oval.	dark.		middle Sept.	Good quality.
Kensington				medium.			no foxiness.		
Kilvington						black.			
Kingsessing									
King		medium.	shouldered	medium.	oval.	white.	sweet, pleas't	middle Sept.	Seedling of Clinton.
King	Sum'er Black								
Kitchen									
Lake's Seedling				large.		black.	sweet.	25th August.	
Large Black Marion									Resembles Isabella, said to be better.
Large White Transparent								very early.	
Laspeyre Grape	Isabella.								
Lebanon Seedling	Catawba.								
Lee's Seedling									
Lenoir		large.	compact.	small.	round.	black.	sweet.	early.	
Little Ozark			long, loose.	medium.		black.		end of Sept.	
Logan		medium.		large.	oval.	black.	sweet.	early.	
Long				medium.		dark purple.	sweet.		Good for wine, not fit for table.
Longworth's Cat'w.				large.					Darker and later than its parent, Catawba.
Longworth's Ohio.	Ohio.								
Louisa		medium.	compact.		round oval	dark purple		14 d. b. I.	

CONTRACTIONS.—(bk.) Black; (s.) Shouldered; (d. b. C.) Days before Catawba; (I.) Isabella.

NAME.	SYNONYM FOR.	BUNCH.		BERRY.		COLOR.	FLAVOR.	RIPENS.	REMARKS.
		SIZE.	SHAPE.	SIZE.	SHAPE.				
Lowell Globe.									
Lowell Globe Fox..	Lowell Globe	small.	....	large.	....	....	....	....	Resembles Charter Oak
Lydia............	....	small.	loose.	medium.	round.	yellowish grn			
Lyman	....	small.	compact.	small.	round.	black.	....	....	Resembles Clinton.
Luffborough	....	....	....	large.	....	purple.	foxy.		
McNeil									
Maddox	....	....	....	....	....	....	....	....	Resembles Norton's Virginia.
Madeira, (of York, Pa.)	Alexander's,								
Mammoth Catawba	....	....	....	....	....	....	....	....	Resembles Catawba, but larger.
Mammoth.........									
Manockanock.....	....	....	....	large.	round.	black.	not sweet.		
Marion	....	large.	....	large.	round.	black.	....	....	Said to be worthless
Marion	Marion Port.								
Marion Port	....	small.	very compact	small.	round.	black.	sweet.	....	Makes excellent wine.
Martha...........									
Martinsburgh									
Massachus'ts White	....	....	....	....	....	....	foxy.		
Miner............	Venango.								
Mary Ann.........									
Mignionette......	....	....	....	....	....	....	....	....	Variety of Vitis Riparia.
Minor............	....	....	....	....	....	....	foxy.		
Minor's Seedling...	Minor.								

Missouri...	...	medium	shouldered	small.	round.	black.	sweet, pleas't		
Missouri Bird's Eye	...	...	long, loose, shouldered	small.	...	black.	sweet.	...	Hardy, makes good wine.
Monteith...	...	...	...	medium.	round.	...	sweet, good.	14 d. b. I.	
Montgomery...	...	...	...	...	...	...	...	...	Resembles Lenoir.
Morin...									
Mottier's White...	...	medium.	compact.	large.	round.	gr'nish white	like Catawba	1st Oct.	Unsuited to the North.
Mottled...	...	...	...	...	...	...	...	...	Resembl's Catawba but the berry is mottled.
Mustang...	...	...	...	small.	round.	black.	pleasant.	...	Makes a good red wine.
Narcissa...	...	...	...	large.	oval.	very dark.	sweet, musky	1 w. b. I.	Seedling of Isabella
Naumkeg...	...	medium.	loose.	medium.	...	clear red.	...	with I.	Seedling of Isabella
North America...	...	...	...	medium.					
North Carolina...	Mary Ann?	small.	loose.	medium.	round.	black.	astringent.	middle Oct.	Good for wine.
Northern Muscadine	...	medium.	loose.	large.	round.	light amber.	foxy, sweet.	10 d. b. I.	Hardy and prolific.
Norton's Virginia...	...	...	long shol'd	small.	round.	dark purple.	brisk, rough.		
Norton's Seedling.	Norton's Vir.								
Oconee...	...	...	...	...	...	...	...	...	Unique, small deep green foliage.
Offer...	...	large.	compact.	large.	...	light brown.	sweet, musky	b. I.	
Ohio...	...	large.	loose.	small.	round.	black.	sweet, brisk.		
Ohio Claret...	...	...	...	small.	...	very dark.			
Ohio Globore...	...	medium.	...	medium.	round.	black.	...	20th Sept.	
Ohio Prolific...	...	small.	...	small.	round.	purple.	acid, vinous.	middle Sept.	
Ontario...	...	large.	...	very large.	round.	black.			
Ozark Favorite...	...	medium.	compact.	medium.	...	...	vinous.		
Osmond...									
Orwigsburgh...									
Pauline...	...	large.	s. compact	medium.	...	brwnsh crim.	sweet.		
Pennell...									
Pennsylvania...									
Perfumed...	Mignionette.								

CONTRACTIONS.—(bk.) Black; (s.) Shouldered; (d. b. C.) Days before Catawba; (I.) Isabella.

NAME.	SYNONYM FOR	BUNCH.		BERRY.		COLOR.	FLAVOR.	RIPENS.	REMARKS.
		SIZE.	SHAPE.	SIZE.	SHAPE.				
Perkins	....	large.	comp't lng	large.	round.	pale red.	sweet, musky		
Perkins	Eudora.	....	....	....	....	....	....	....	Resembles Isabella.
Pioneer	....	....	....	....	....	....	....	....	Seedling of Mammoth Catawba.
Pæschel's Mamm'h	....	....	....						
Pond's Seedling	....	....	long.	small.	round.	blue.	pleasant.		
Potter		....	....	large.	round.	dark amber.	sweet, foxy.	....	Hardy and vigorous
Potter's Catawba	Potter.								
Powell	Bland.			medium.	....	pale red.	sweet, foxy.		
Purdy's Prolific	....	....	....			purplish.	....	....	Same as Catawba but earlier.
Purple Catawba	....	....	....	....	....				
Raabe	....	small.	shouldered	small.	round.	purple.	sweet, vinous	....	Probably hybrid between Catawba and Edinburgh.
Raabe's No. 3.	....	small.	compact.	small.	round.	dark red.	sacch'n, foxy		
Raabe's No. 4.	Clara.								
Raabe's Clara	Clara.								
Ramsdell	....	....	....	large.	round.	dull red.	juicy, sweet.	....	Hardy, does best on light soil.
Ranney									
Rathrack									
Rebecca	....	medium.	compact.	medium.	obovate.	gr'nish white	sweet, musky		
Red Muncy	Catawba.								
Red Scuppernong	Bland.								
Roanoake	Scuppernong								
Sage	....	medium.	....	large.	round.	dark red.	foxy.	1st Sept.	Hardy.
St. Catherine	....	....	loose.	large.	....	red.	sub-acid, f'xy		

Saluda									Resembles Isabella.
Scuppernong		small.	loose.	large.	round.	black.	sweet, musky		
Scuppernong		small.	loose.	large.	round.	white.	sweet, musky		
Schuyler				large.	round.		acid.	middle Sept.	
Segar Box	Ohio.								
Shepard		medium.	shouldered		round.	pale red.	sweet, musky		
Shaker Grape	Union Vill'ge								
Shurtleff		medium.	shouldered	small.	round.	black.	a., sprightly.	10 d. b. I.	
Sherman									
Schuylkill Muscadell	Alexander's, Canby's Aug of some.								
Schuylkill Muscadine	Alexander's, Canby's Aug of some.								
Singleton	Catawba.								
Smart's Elsinburgh.	Elsinburgh.								
Spring Mill Constantia	Alexander's, Canby's Aug of some.								
Strawberry									
Summer Black		small.		medium.	round.	black.	sweet, foxy.	end of Aug.	
Sumpter	Lenoir.								
Suntara		small.		small.		black.	sweet.	very early.	
Tasker's Grape	Alexander's.								
Taylor	Bullitt.								
Taft		medium.	compact.	large.	globular.	light green.	sweet, foxy.		
Texas Blue Muscadine							good flavor.	early.	
Texas Post Oak									Foliage resembles Post Oak.
Texas Sand Hill									
Thurmond	Herbemont?								
To Kalon		large.	shouldered	large.	oval.	black.	sweet, lusc'us	10 d. b. I.	
Transparent				large.	round.	red.	acid.	early.	
Troy				large.	oval.	dark purple.	sprightly, a.		
Troy Grape	Troy.								
Tryon									
Union Village		very large.	compact.	very large.	round.	black.	juicy, good.	1 w. b. I.	Very vigorous.
Venango			compact.	large.	round.	brown'h lilac			

CONTRACTIONS.—(v.) Vinous; (a.) Acid; (s.) shouldered; (d. b. I.) Days before Isabella.

* NAME.	SYNONYM FOR.	BUNCH.		BERRY.		COLOR.	FLAVOR.	RIPENS.	REMARKS.
		SIZE.	SHAPE.	SIZE.	SHAPE.				
Vernet Grape	Isabella.								
Vine Arbor				large.	round.	black.	sweet.	late.	
Virginian Amber				small.		pale red.			
Wachita									
Warren		large.	long, loose.	small.		blue.	brisk.		
Warren's Catawba									Seedli'g of Catawba said to ripen early
Warrenton	Herbemont.								
White Catawba									Same as Catawba, but white.
Wilcox				small.		gr'nish white	acid.		
Wilmington				medium.	oval.	white.	high flavor'd		
Willington									
Wine Horne			long, loose.	medium.		black.	vinous.		
Winne	Canby's Aug								
Winslow									Resembles Clinton.
Wintermoot					round.	black.	juicy.		
Winton			loose.	small.	round.		sweet, vinous	early.	Hardy.
Woodson			long.	medium.	round.	chocolate.	acid, vinous.	end Sept.	Good for wine, unfit for table.
Worthington	Clinton?								Describ'd by Adlum
Wright's Isabella									Resemb's Franklin, good wine grape.
Wyoming				medium.		black.			
York Madeira									Said to resemble Miller's Burgundy
Young's Seedling									Seedling of Isab'lla, said to be earlier.
Zane									

THE MANUFACTURE OF WINE.

It is still a disputed question whether or not it is possible for good wine to be manufactured in the United States. Daniel Webster, whose high intellectuality did not detract from his fondness for the pleasures of the table, declared that we could never hope to make good wine on this continent, and that it would always pay us better to raise corn, cotton, etc., for export, and buy our wines and silks. On the other hand, the following letters from President Jefferson to Mr. Adlum would seem to establish the fact that, even at an early day, wine had been made in this country of more than ordinary quality:

EXTRACTS OF LETTERS FROM MR. JEFFERSON, LATE PRESIDENT OF THE UNITED STATES.

Dated October 7th, 1809.

"While I lived in Washington, a member of Congress from your State (I do not recollect which) presented me with two bottles of wine made by you, one of which, of Madeira color, he said was entirely factitious; the other, a dark red wine, made from a wild or native grape, called in Maryland a Fox grape, but very different from what is called by that name in Virginia. *This was a very fine wine, and so exactly resembling the red Burgundy of Chamberlin (one of the best crops) that on a fair comparison with that, of which I had very good on the same table, imported by myself from the place where made, the company could not distinguish the one from the other.* I think it would be well to push the culture of that grape, without losing our time and efforts in search of foreign vines, which it will take centuries to adapt to our soil and climate."

Dated April 20th, 1819.

"The quality of the bottle you sent me before satisfies me that we have at length found one native grape inured to all the accidents of our climate, which will give us a wine worthy the best vineyards of France. When you did me the favor of sending me the former bottle, I placed it on the table with some of the *best Burgundy of Chamberlin, which I had imported myself from the maker of it,* and desiring the company to point out which was the *American bottle,* it was acknowledged they could perceive no difference."

Dated April 11, 1823.

"I received successively two bottles of wine you were so kind as to send me; the first, called Tokay, *is truly a fine wine,* of high flavor, and as you assure me there was not a drop of brandy or other spirit added to it, I may say it is a wine of a good body of its own. The second bottle, a red wine, I tried when I had good judges at the table; we agreed it was a wine *one might always drink with satisfaction,* but of no peculiar excellence. Speaking of brandy being added to the wine, he says it is never done but by the exporting merchants, and then only for the English and American markets, where, by a vitiated taste, the intoxicating quality of wine, more than its flavor, is required by the palate."

Now Mr. Jefferson and his friends were no doubt accustomed to drink good wines, and we think their opinions valuable, although at the same time it must be confessed that they were not very extraordinary judges, or they would have detected a *difference* between the French and American wines. The question of *superiority* may sometimes be disputed even by good judges, that of identity never.

Good wine has also been made in the south of England, as the following extract from Barry's work on wines will show, and as it contains some practical notes on wine-making, we give it entire:

"The vineyard of Painshill is situated on the south side of a gentle hill; the soil a gravelly sand; it is planted entirely with the two sorts of Burgundy grapes: the Auvernat, which is the most delicate, but the tenderest; and the Miller grape, commonly called the black cluster, which is more hardy. The first year I attempted to make red wine in the usual way, by treading the grapes, then letting them ferment in a vat till the hulls and impurities formed a thick crust at the top, the boiling ceased, and the clear wine was drawn off from the bottom.

"This essay did not answer; the wine was so very harsh and austere, that I despaired of ever making red wine fit to drink. But through that hardness I perceived a flavor something like some small French white wines, which made me hope I should succeed better with white wine. That experiment succeeded far beyond my most sanguine expectations; for the very first year I made white wine, it nearly resembled the flavor of champagne, and in two or three years more, as the vines grew stronger, to my great amazement, my wine had a better flavor than the best champagne I ever tasted. The first running was as clear as spirits, the second running was *œil de perdrix*, and both of them sparkled and creamed in the glass like champagne. It would be endless to mention how many good judges of wine were deceived by my wine, and thought it superior to any champagne they ever drank; even the Duke de Mirepoix preferred it to any other wine. But such is the prejudice of most people against anything of English growth, I generally found it most prudent not to declare where it grew till after they had passed their verdict on it. The surest proof I can give of its excellence is that I have sold it to wine merchants for fifty guineas a hogshead; and one wine merchant, to whom I sold five hundred pounds' worth at one time, assured me he sold some of the best of it from 7s. 6d. to 10s. 6d. per bottle.

"After many years' experience, the best method I found of making and managing it was this: I let the grapes hang till they got all the maturity the season would give them. Then they were carefully cut off with scissors and brought home to the vine barn in small quantities, to prevent their heating or pressing one another; then they were all picked off the stalks, and all the moldy or green ones were discarded before they were put upon the press, where they were all pressed in a few hours after they were gathered; much would run from them before the press squeezed them, from their own weight one upon another. This running was as clear as water and sweet as syrup, and all this of the first pressing, and part of the second, continued white. The other pressings grew reddish, and were not mixed with the best. As fast as the wine ran from the press into a large receiver, it was put into hogsheads and closely bunged up. In a few hours one could hear the fermentation commence, which would soon burst the casks if not guarded against by hooping them strongly with iron and securing them in strong wooden frames and the heads with wedges. In the height of the fermentation I have frequently seen the wine oozing through the pores of the staves.

"These hogsheads were left all the depth of winter in the cool barn to reap the benefits of the frosts. When the fermentation was over, which was easily discovered by the cessation of noise and oozing—but, to be more certain, by pegging the cask—when it would be quite clear, then it was racked off into clean hogsheads and carried to the vaults, before any warmth of weather could raise a second fermentation. In March the hogsheads were examined. If they were not quite fine, they were fined down with common fish glue, in the usual manner; those that were fine of themselves were not fined down, and all were bottled about the end of March, and in about six weeks more would be in perfect order for drinking, and would be in their prime for above one year; but the second year the flavor and sweetness would abate and would gradually decline, till at last it lost all flavor and sweetness, and some that I kept sixteen years became so like old hock that it might pass for such to one who was not a perfect connoisseur. The only art I ever used to it was putting three pounds of white sugar-candy to some of the hogsheads, when the wine was first tunned from the press, in order to conform to a rage that prevailed to drink none but very sweet champagne.

"I am convinced that much good wine might be made in many parts of the south of England. Many parts are south of Painshill, many soils may be yet fitter for it, and many situations must be so, for mine was much exposed to southwest winds (the worst of all for vines) and the declivity was rather too steep. Yet with these disadvantages it succeeded many years. Indeed, the uncertainty of our climate is against it, and many fine crops have been spoiled by May frosts and wet summers. But one good year balances many disappointments.

"Captain St. Pierre, who has established a great colony of vignerons in South Carolina, and carried there three years ago above three hundred vignerons from different parts of Europe, was with me several days before his departure, was charmed with my vineyard, and he had cultivated vineyards many years in France. He was very happy at my giving him all the cuttings of my vineyard, as he found it very difficult getting the right sort, and though his plantations are about the latitude of 33°, he has not the least doubt of having excellent wine there, which, if he has, must be of infinite service to this country."

Still more recently Mr. Longworth has succeeded in the manufacture of fine champagne wines, which we believe are valued as high as any, except the very finest brands of foreign wines.

Wine is the fermented juice of the grape, and *pure wine* should contain nothing else. When sugar and spices are added, and exist in the fluid as such, the product is no longer *wine*, but *liqueur* or cordial. Some have, however, extended this principle so far as to assert that any addition to the juice of the grape, either before or after its fermentation, robs it of its claim to the name of wine; but to this we cannot subscribe. If we by any process could produce a fluid identical in its chemical and physical properties with the juice of the grape, we could no doubt make good and real wine therefrom. And if so, then surely the addition of any ingredient which may be required to bring the juice up to the quality and composition of a good wine-making must, cannot have any but a good effect, and must produce a real wine.

Now the juice of the grape varies in composition from several causes. The variety of grape, the climate in which it is produced, the character of the soil in which it grows, the nature of the manure with which it has been nourished, the mode in which it has been pruned, its exposure to sun and air, and many other influences, all modify the character of the must, and consequently of the wine produced therefrom. In almost every locality we are confined to a few varieties of grapes, and as the climatic conditions are also in a great measure beyond our control, we must depend upon judicious pruning, manuring and cultivation for the production of the best grapes for the manufacture of wine. In former chapters we have detailed the peculiarities of vine-dressing as adapted to the producing of wine-making grapes;

but we may be excused for briefly recapitulating them.

Must for wine requires to be highly saccharine, and although the wines manufactured from American grapes have not yet shown much inorganic matter (potash salts) in their composition, yet the best wines in Europe are made from grapes containing an extra quantity of these matters. In order, therefore, to the production of a good wine, it will be requisite to produce grapes not only thoroughly ripened by A HOT SUN ACTING ON THE LEAVES, but they should also contain the juices and inorganic salts in large amount.

With a view to this, it will be necessary in the fall, and shortly after the vintage, to lightly fork in a dressing of bone-dust, guano or hen manure; and on the fall of the leaf, and before any frosts set in, the border should be covered with the fallen leaves raked together and mixed with stable litter or cleanings. This will protect the roots from the severity of our winters, and enable them to sustain the draft made in spring by the branches at an earlier date than they otherwise would.

In the spring, after the weather has become settled, the border should be very lightly forked over and the long litter removed; the rest may be mixed with the surface soil.

The vine having been properly pruned, must be allowed to *break its buds*, as it is termed, and push out the young stems until those which promise best can be clearly distinguished.

As soon as the leaves are formed, liquid manure may be applied if the number of vines cultivated will permit of it, and this application of liquid manure may be continued until after the middle of July. It should then cease for the season. Meanwhile, as soon as the young shoots are well formed, all the weakly ones should be rubbed off, carrying the pruning recommended in former pages to even a greater degree of severity than there noted.

By these means the grapes will be obtained ripe much earlier and of a higher (not stronger) flavor. The importance of having the grapes ripe early will be appreciated when we consider that, other things being equal, the heat and dryness of the season in which they ripen will be the measure of the perfection of the grapes, at least in this latitude. Now, in 1858, the mean temperature of August was 69° Fahrenheit, while the mean temperature of September was only 61°, and as the amount of rain which fell in each month was equal, the grapes which were ripe by the beginning and middle of September were much richer in saccharine and other wine-making elements than those which were produced in the cool and damp atmosphere of September and October.

From the foregoing observations it will be evident that in preparing must for wine we must pay particular attention to the quality of the grapes and the circumstances under which they were raised. Thus, in Cincinnati, no sugar is added to the juice of the Catawba; it is fermented just as it comes from the press. But in more northern climes, not only does the juice of the Isabella and Clinton require sugar, but that of the Catawba stands in need of it, in order to make, not a sweet but a full-bodied wine, which will bear keeping.

In the manufacture of wine from the grape, the first process is to carefully pick over all the grapes, rejecting those which are unripe, rotten, mildewed, or imperfect in any other way. The rejection of the stems will depend upon the character of the wine desired. If retained, they impart a roughness to the wine, which some admire; and it is claimed by some, that the tannin of the stems helps to preserve the wine. The grapes are then to be mashed, which is easily done with the hands if in small quantity. In the large way it is performed by passing the grapes between rollers armed with pins. On a smaller scale, a beetle or stamper, armed with pins, may be used; and where but a few are prepared (as for domestic purposes) the hands alone can perform the work. A gentleman of this city has devised a very useful and efficient machine, in which, by passing the grapes between rollers covered with india-rubber, the juice is expressed and separated from the husks without bruising the stems or seeds.

If prepared in the ordinary way, the must may be allowed to ferment either before or after the juice has been separated from the seeds and husks. Fermenting the husks and seeds gives a roughness and harshness to the wine as well as a higher color. For the finest wines the juice only is fermented.

This is effected by simply allowing the juice to

stand in casks filled three-fourths full. Fermentation speedily sets in; the saccharine matter becomes converted into carbonic acid, which escapes, and alcohol, which remains in combination with the fluid, and gives it the character of wine. At first the fermentation is very violent, but after a time it moderates, when the casks should be filled up, lightly bunged, and kept during winter in a temperately cool apartment. In spring it should be carefully drawn off, either by means of a syphon or through a hole bored into the cask some distance above the bottom, so as to avoid disturbing the lees. After this, fermentation should be avoided as much as possible, which is best effected by a low temperature and the exclusion of oxygen. It is generally considered best, we believe, to leave the wine at least one season in the cask into which it has been drawn off. In some cases it is kept for years in the "wood," as it is termed.

Wine can of course be made of any kind of grape, though in and around Cincinnati the Catawba is altogether preferred. Tolerable wine has been made of the Isabella, and in the hands of Dr. Underhill it has proved of superior excellence for this purpose. But for all northern localities we think the Clinton promises to be the wine grape. When carefully pruned and *thinned*, so as to get fair bunches instead of the load of little sour trash usually seen, the Clinton grape is peculiarly rich in saccharine and saline matter. Of its wine-making qualities Nicholas Longworth speaks as follows in a letter to "The Horticulturist:"

"I believe I advised you that the must and wine of the Clinton grape differed from any I have ever seen. The must weighs very heavy, indicating a large quantity of saccharine matter; the wine, fully fermented, acid and weighing but little, and indicating but little spirit. Of the grapes you sent last spring I made two kinds of wine. One part I pressed as soon as worked, and put at the rate of seventeen ounces of sugar to the gallon of must; the other I worked and left to ferment in the skins before pressing, and put no sugar. The first is a beautiful dark red, which I have never seen equalled, and very clear. It has no sweetness and is rather dry, but of fine flavor. The other is clear, very dark red, and more acid, but of fine flavor. I deem that in our warmer latitude the must will have more sugar, and will make a valuable red wine, an article we have not at present.

"I am very desirous of giving the grape a further trial, and shall esteem it a favor if you will engage and send me from two to five bushels of grapes, and let them be as ripe as possible. I shall also be pleased to get from two to five thousand cuttings. I will next spring graft a dozen roots with this grape, and the next season guarantee to have grapes enough to test how they will suit our climate, as I have had grafts grow the first season from ten to thirty feet, and often bear some fruit the same season."

The following letter, received from a lady whose wine we can testify to be of very superior excellence, contains directions slightly different from those in ordinary use, and in some respects perhaps superior. We give it in her own words, which it may be but justice to say, were not originally intended for publication:

"After the grapes are gathered, pick carefully from the clusters all the good ones. Wash these, being careful not to mash the seeds (we had a little machine for this purpose that turned with a crank). Have ready a perfectly sweet cask, that has a hole, about an inch in diameter, bored in one side near the bottom; fit into this hole a stick from six to eight inches long, with a hole bored from end to end of sufficient size to let the juice flow freely through it. Stop this hole tightly with a plug; as the grapes are mashed, pour the juice, skins, pulp and all, into the cask. When all are in, cover closely with four or five thicknesses of woollen blankets; let it remain in this condition until fermentation has advanced sufficiently to cause the grapes or must (as I believe wine-makers call it) to rise to the top and begin to crack open, the cracks being filled with little yeasty-like bubbles, which will be probably in from four to eight or ten days, according to the temperature of the weather. Now have ready a perfectly clean barrel, purified with sulphur; put into a pail ten or twelve pounds of sugar, take out the little plug, and let the juice on the sugar. As you fill the pail, stir the sugar occasionally from the bottom, so as to dissolve enough of it to make the juice sufficiently sweet. If the sugar should all dissolve before the juice is all drawn out, of course put in more. When the barrel is full, put the bung in lightly, so as to give it a chance to ferment. The little cups you speak of were used more as an experiment than a necessity; when those were used, the bung was fitted in tight and a small hole made in the bung, and a tin tube inserted in it, rising from the bung, the long end being in the bung, and the short end in a little tin cup filled, and kept full of water, care being taken to keep the barrel always full; but, as I said before, this was not necessary. After the juice had been barrelled, as above described, let it stand till some clear, cold day in February. Then draw off the juice and put it in another barrel, care being taken to have it perfectly clean and well fumigated as the first was; save a pailful, and when all has been drawn off, stir into this pailful the whites of ten or twelve eggs, beaten to a froth, as you would for cake. When well stirred, pour this in the barrel with the rest. After being well incorporated with that in the barrel, bung it up tightly, and for two years 'touch not, taste not, handle not,' and as much longer as you can resist the temptation, as it improves from 25 to 50 per cent. in quality every year it is suffered to stand. The barrels should be kept in a dark cellar.

"The above contains all the most important particulars of the doctor's process of making wine, to the best of my recollection. It will answer very well where one only desires to make a little for his own use; but would hardly answer on a large scale.

"Fumigating the barrels with a sulphur match destroys any musty or unpleasant smell which the barrel may have, and is done by melting flowers of sulphur or roll brimstone in an iron vessel on the stove; making a swab by rolling a rag around the end of an iron rod, saturate the rag with the melted sulphur as you roll it around; stick the other end of the rod into a good sized potatoe, and set fire to the rag or swab; hang it in the barrel at the bung-hole, the potatoe will prevent it dropping down in the barrel."

The following recipes for currant wine are perhaps more useful than appropriate. They have been collected from various reliable sources, and it is probable that few will regret their insertion

CURRANT WINE.*

Three varieties of currants are employed in making wine—white, red and black; but the two first are most common. The wines from the white and red sorts differ a little from each other in color, also in flavor. With proper management they are capable of producing a wine analogous to the lighter wines of the grape, according to Dr. MacColloch, "not easy to be distinguished from the Colares of Portugal, which although not in the first class, is certainly superior to most of our domestic wines." A principal defect in currant wine, as commonly made,

* Copied from Webster's "Encyclopædia of Domestic Economy."

arises from too small a quantity of the fruit being used, and of course too much sugar and water. On this account, and from the imperfect fermentation, these wines are usually too sweet; and from a natural bad flavor in the husks, which are often kept in the must, a mawkish taste is introduced. By increasing the quantity of the fruit, which is generally used only in the same proportion as in gooseberry wine, and avoiding the use of the husks, the flavor and quality of the wine are materially improved.

At present only sweet wines are generally made from currants; but dry wines may also be fabricated from this fruit by the method already pointed out; for these the fruit should be ripe.

Brisk wine may also be made, and then a proportion of unripe fruit should be introduced. The use of tartar, likewise, Dr. MacColloch is of opinion, would be advantageous, and would correct a defect not uncommon, that of having an ammoniacal taste. Another improvement has been put in practice with success, not only in making currant wine, but in all those wines produced from fruits of which the flavor is either bad or which have little or no flavor; this is by boiling the fruit juice previously to fermentation. From this treatment many tasteless fruits acquire a flavor, and many bad flavors are converted into agreeable ones. This is particularly remarkable in the case of the black currant, which, though harsh in its natural state, acquires by boiling a powerful and to most persons an agreeable flavor. Wine made from this fruit in a raw state has no particular property, whereas that of the boiled may be, by careful management, brought to resemble some of the best of the sweet Cape wines. The boiling must not be too long continued, as this degree of heat tends to coagulate and precipitate the ferment, and thus render it ineffective. Some artificial ferment is generally necessary with boiled juice. Great care must be taken in separating the stalks, and if the skins and solid matter are fermented in the vat, they must not, at all events, be introduced into the casks. Many persons put the pure juice into the casks at once, strained, without any previous fermentation in the vat.

RECIPE FOR WHITE CURRANT WINE.

Bruise forty pounds of the fruit in a tub of the capacity of fifteen or twenty gallons, and add to it four gallons of water. Stir the whole well, and squeeze till the pulp is thoroughly separated from the skins; leave these materials at rest for about twelve hours, and then strain them through a canvas bag or fine hair sieve, and pass one gallon of fresh water through the marc. Dissolve thirty or twenty-five pounds of white sugar in the juice thus obtained, and make up the whole quantity by an addition of ten gallons and a half of water. The proportion of sugar here given is for a brisk wine; if a sweet wine is required, there must be forty pounds of sugar. White sugar is recommended as much the best. If moist sugar be used, somewhat more will be necessary. The must being now prepared, the fermentation and subsequent treatment must be exactly the same as for gooseberry wine, and the reader may therefore refer to that recipe.

If brandy is to be added, it should be added toward the end of the fermentation in the cask. For the above quantity some will put in a quart of brandy alone; others mix it with honey.

Whether the wine should be racked off from the lees at the end of six months, put into a cask for six months longer before it is bottled, or be suffered to remain the whole time in the lees, must depend upon the state of the wine according to the principles explained above. The bottling should be carefully attended to.

ANOTHER RECIPE.

White currants, nine gallons; white gooseberries, one gallon; white sugar, twenty-five pounds; white tartar, an ounce; bitter almonds, two ounces; water, nine gallons; brandy, one gallon.

MR. CORNELL'S RECIPE FOR MAKING RED CURRANT WINE.

Bruise eight gallons of red currants with one quart of raspberries. Press out the juice, and to the residuum, after pressure, add eleven gallons of cold water. Add two pounds of beet-root, sliced as thin as possible, to give color, and let them infuse, with frequent mixture, for twelve hours; then press out the liquor as before, and add it to the juice. Next dissolve twenty pounds of raw sugar in the mixed liquor, and three ounces of red tartar in fine powder. In some hours the fermentation will commence, which is to be managed according to the details for gooseberry wine and the principles we have stated previously. When the fermentation is completely over, add one gallon of brandy; let the wine stand for a week, then rack off, and let stand for two months. It may now be finally racked off, bunged up in a cask, and set by in a cool cellar for as many years as may be required to ameliorate it.

BLACK CURRANT WINE

May be made in the same manner, using six gallons of black currants, three gallons of strawberries, twenty-five pounds of raw sugar, four ounces of red tartar, ten gallons of cold water, and three quarts of brandy.

ELDERBERRY WINE.

The elderberry is well adapted to the production of wine. Its juice contains a considerable portion of the principle necessary for a vigorous fermentation, and its beautiful color communicates a rich tint to the wine made from it. It is, however, deficient in sweetness, and therefore demands an addition of sugar. There are several methods of making this wine; the following are some of the most approved recipes:

Take one gallon of ripe elderberries and one quart of damsons or sloes, for two gallons of wine to be made; boil the fruit in about half the quantity of water till they burst, breaking them frequently with a stick. Strain the liquor and return it to the copper. To produce eighteen gallons of wine, twenty gallons of this liquor are necessary, and for whatever quantity the liquor falls short of this, water must be added to make up. Boil this, together with fifty-six pounds of coarse moist sugar, for half an hour, and it is to be fermented in the usual manner when sufficiently cool, and then is to be tunned or put into the cask. Put now into a muslin bag a pound and a half of ginger, bruised, a pound of allspice, two ounces of cinnamon, and four or six ounces of hops; suspend the bag with the spice in the cask by a string, not long enough to let it touch the bottom; let the liquor work in the cask for a fortnight, and fill up in the usual manner. The wine will be fit to tap in two months, and is not improved by keeping like many other wines. Elderberries alone may be used.

ANOTHER METHOD.

Elderberries, ten gallons; water, ten gallons; white sugar, forty-five pounds; red tartar, eight ounces; fermented with yeast in the usual manner. When in the cask, ginger root, sliced, or allspice, four ounces; bitter almonds, three ounces; suspended in a bag, may be allowed to infuse in the liquor when it is fermenting; they are then to be removed. Brandy may be added or not. When the wine is clear, which will be in about three months, it may be drawn off from the lees and bottled. The spices may be varied according to taste.

BLACKBERRY WINE.

To one quart of juice two quarts of water and three pounds of sugar. The berries to be mashed cold, and the juice expressed and strained. The sugar dissolved in the water and strained. The whole then mixed in kegs and placed in a cool cellar. The bung-hole to be left open until fermentation has nearly ceased, then closed tight and left standing until the ensuing April, when it should be carefully drawn and bottled.

STRAWBERRY OR RASPBERRY WINE.

Bruise and press out the juice of either fruits; pour on the marc seven gallons of water; infuse for twelve hours and press out the liquor. Add this liquor to the juice, and mix them with six gallons of cider. Dissolve in the mixture sixteen pounds of raw sugar and three ounces of powdered red tartar, and then set it to ferment in the usual manner. Pare the rinds of two lemons and of two oranges, and together with the juice throw them into the fermenting tub, and take out the rinds when the fermentation is over. Three gallons of brandy may be added. In making raspberry wine, a gallon of white and red currant juice should be added, and an equal quantity of water left out.

ORANGE WINE.

Seville oranges are used for this purpose; they are best in March. For eighteen gallons of wine half a chest of oranges are required. Pare the rinds from about a dozen, or two dozen, as more or less of the bitter will be agreeable. Pour over this a quart or two of boiling water, and after letting this stand for twelve hours, strain off the water, which extracted much of the essential of the oranges. Take the peel off entirely from the remainder of the oranges, squeeze the juice through a bag or sieve, and put it into a cask with about forty-five pounds of white sugar or fifty-five of the best moist sugar. Soak the pulp in water for twenty-four hours, and after straining this, add it to the cask. Repeat this several times till the cask is full. Stir the whole well with a stick till the sugar is dissolved, then set it to ferment. The fermentation is slower than with currant wine, but may be heard hissing for several weeks. When this subsides, close the bung-hole, and proceed as in the case with gooseberry wine. Some add brandy. The wine requires to be kept in the cask a year before it is bottled.

GINGER WINE.

Dissolve eighteen or twenty pounds of sugar in nine and a half gallons of boiling water, and add to it ten or twelve ounces of bruised ginger root. Boil the mixture for about a quarter of an hour, and when nearly cold add to it half a pint of yeast, and pour it into a cask to ferment, taking care to fill the cask from time to time with the surplus of the liquor made for that purpose. When the fermentation ceases, rack off the wine, and bottle it when transparent. It is a common practice to boil the outer rind of a few lemons together with the ginger destined for the wine, to impart to the wine the flavor of lemon peel.

CURRANT WINE.

Gather the currants when fully ripe; press and measure the juice; add two-thirds water, and to each gallon of that mixture put three pounds of Muscovado sugar (the cleaner and drier the better; very coarse sugar, first clarified, will do equally well); stir it well until the sugar is quite dissolved, and then tun it up. Do not let the juice stand over night before mixing; or at least not so long as to ferment.

Make rather more than to fill the casks, so as to fill them up after drawing off the wine.

Lay the bung lightly on the hole, to prevent flies, etc., from creeping in. In three weeks or a month after making, the bung-hole may be stopped up, leaving only the vent-hole open, until the wine has done working, which will be about the latter end of October. It may then be racked off into other clean casks; but some persons prefer letting it stand on the lees until spring, as it thus acquires a stronger body and is in a great measure divested of that sweet, luscious taste peculiar to made-wine. It may without damage stand two years on the lees.

When it is to be drawn off, bore a hole at least an inch above the tap-hole, a little to the side of it, that it may run clear off the lees.

APPENDIX I.

EXAMPLES OF AMERICAN VINEYARD PRACTICE.

THE OHIO SYSTEM OF VINEYARD CULTURE.

THIS is merely a modification of the French and German methods, having been generally introduced by vine-dressers from those countries. It is, we believe, now generally giving place to the trellis system of culture, which seems to be better adapted to the habit of our native vines. Vines and even vineyards may be found around Cincinnati, which are trained differently from the method here described, but nevertheless, the following is what is known as the *Ohio system.*

The ground having been properly prepared, the vineyard is set out either with cuttings or rooted plants, generally the former. In setting out cuttings, holes about two feet deep are made with a stilt or dibble, shod with iron, and after inserting two cuttings in each, the holes are filled in with sand which is washed into immediate contact with the cuttings by means of water. During the first season, the vines are allowed to grow at random, the ground, however, being kept clean and mellow.

In the spring of the second season the vines are pruned, which is done by removing all the wood made by the young cutting, and also all the roots which spring from the cutting, within several inches of the surface. Fig. 1 shows the young plant. The soil being removed, the roots *e*, *e*, *e* are cut off close to the

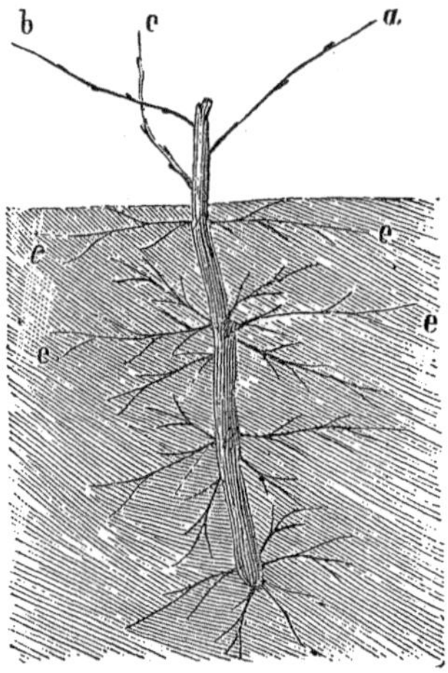

Fig. 65.

stem, the shoots *a b* are cut clean out, and *c* is cut down to one eye, which should be as near the old wood as possible, and if on it, so much the better. During the second year the vines are treated nearly the same as the cuttings were during the first year, and the spring pruning is also the same.

During the third summer, three or four shoots are trained up and carefully tied to stakes; laterals are pinched out and the shoots stopped in September.

During the fourth year, the vines are allowed to bear on the spurs produced by cutting back the shoots of the previous season to six or eight inches. These spurs of course throw up fruit-bearing canes, which during the fifth season are tied to stakes in bows, so as produce a crop of grapes, and at the winter pruning the bows are cut away, their place being filled next season by a fresh cane trained up for the purpose during the preceding summer.

The following figures will illustrate this fully: Fig. 66 shows the vine in the fall of the fourth year; H is the head of the

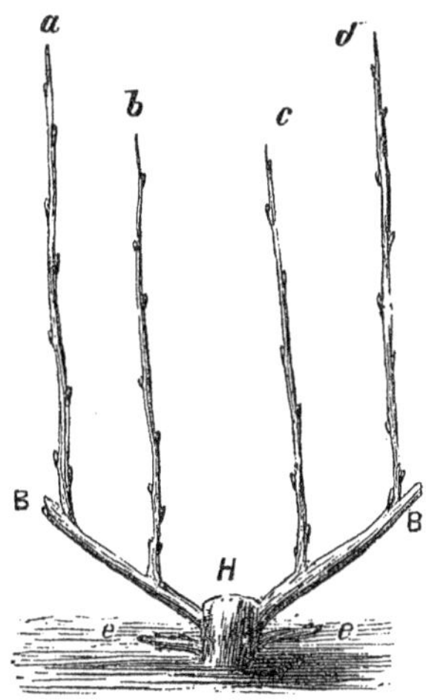

Fig. 66.

vine, B the arms or thighs, as they are sometimes called; and *a*, *b*, *c*, *d* are the canes which bore fruit last year; *b* and *c* are cut off to one good bud, and *a* and *d*, after being shortened, are formed into bows and tied to stakes, so that the vine in the spring of the fifth year presents the appearance shown in Fig. 67. The bow will now yield a liberal crop of grapes, and a few bunches will be obtained from the shoots springing from the spurs *b* and *c*, though they must not be allowed to bear much, as it is desired that they should grow strong and vigorous so as to form the bows for next year. If the vines are strong, they may be allowed to bear more, and other spurs are sometimes allowed to grow from the arms where the vines will bear it.

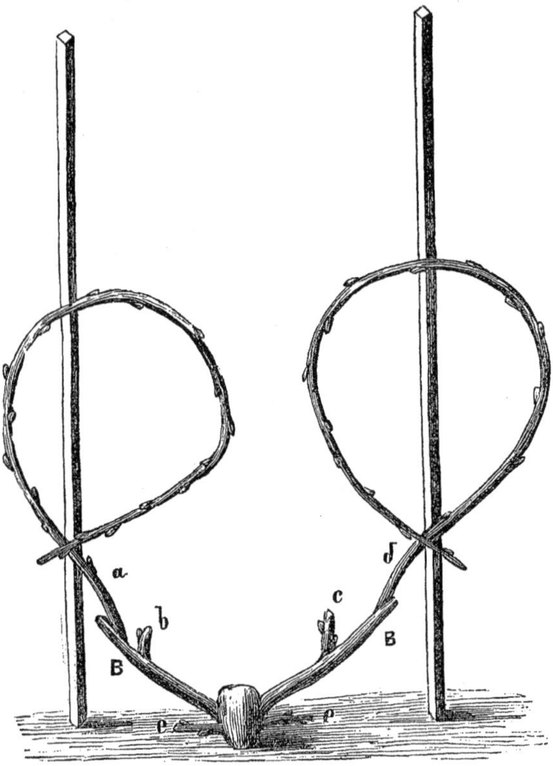

Fig. 67.

The arms themselves are renewed every few years, so as to get rid of all the old gnarled spurs, by training new shoots from the spurs *e e*.

DR. UNDERHILL'S VINEYARDS AT CROTON POINT.

The following account of Dr. Underhill's Vineyards is taken from the "Country Gentleman" of September 25th, 1856. Since that account was published, Dr. Underhill has greatly extended his vineyards, and is thus enabled to devote more of his grapes to the production of wine without lessening the quantity of fruit sent to New York market.

"The readers of our papers have long been familiar with the name of Dr. Underhill as a grower of Isabella and Catawba vines, and lovers of well ripened and carefully marketed grapes in New York city, as the most extensive producer of this fruit in its vicinity. He began to plant the varieties named, or at least the former of them, about twenty-five years ago, having previously made some unsuccessful attempts at growing foreign sorts without shelter; and he has been untiring in his subsequent efforts to attain the best mode of cultivation in every particular, from the first setting of the slip, to the productive maturity of the plant in the vineyard. He is now in possession of nearly a hundred acres of land, of which upward of forty are in grapes, or, with the addition of adjoining vineyards belonging to his brother, there are more than fifty acres in all, to the sale of plants and the marketing of fruit from which Dr. Underhill gives his undivided attention.

"Croton, or Teller's Point, as it was formerly called, juts into the river fully half its width, dividing Haverstraw bay above from the Tappan Zee below. The stream from which it has received the name it now generally goes by, falls into the Hudson on the south—what is left of it after being dammed and drained off for the benefit of New York city. The extreme point of the little peninsula turns downward, commanding in this direction one of the finest river views among the many beautiful ones for which the Hudson is justly famous. Here, once in Revolutionary times, was fired a humble cannon at the *Vulture* in the bay below—scaring her from her anchorage, and leaving Andre without means of safe escape from the plot he was projecting with the traitorous Arnold. The soil is nearly a pure gravelly sand, underlaid at a depth of twenty or thirty feet with clay, and bordered here and there at the river's edge with alluvial deposits. Occasionally the upland is slightly loamy, but for the most part entirely sand, as above described.

"Dr. Underhill plants his vineyards either in spring or fall as may be convenient, setting the vines seven feet apart, in rows six and a half feet from each other. This will take about one thousand to the acre. In his position as to climate and weather, he thinks the question of the inclination of the land immaterial, though further north he would prefer an eastern or southern exposure, or one varying near these points. He has found it best to place the rows so that the prevailing summer winds may have free course through them—contrary to the European practice, in which circulation of the atmosphere is avoided, chiefly on account of the frequency of cold storms. He has found that here it is beneficial, preventing mildew and promoting the healthiness of plant and fruit.

"In the number of plants to the acre his practice is also widely divergent from that in Germany and about Cincinnati—where twenty-five hundred is an ordinary thing. By placing them at greater distances he is enabled to 'secure a crop the first year,' as he remarked—if not of grapes, of *something else* between the rows, and as the vines do not bear until the third summer this is a matter of some importance. They are also taken care of much more easily, as horses can be employed to cultivate the ground, where only men could otherwise be admitted, and, finally, he thinks the yield quite as good and great, as can be produced from more plants on the same space. In fact, in ten years, if the vines crowd at all, or the land is too rich, he sometimes finds it expedient to remove every other vine in the rows, thus leaving only five hundred to the acre. One man, according to his mode, cares for six acres—at least four times as much as he could do on the German plan. Dr. Underhill is opposed on the most stringent principles to allowing any of his land to lie waste and idle, and by obtaining two crops from it before the grape becomes large enough to produce, compels the vineyard to pay while it is being made, though after the vines begin to yield he entirely excludes every other species of vegetation.

"To adapt the ground as nearly as possible to the exact wants of the grape, has been the subject of many and long experiments with Dr. Underhill. During his first trials he expended a great deal on artificial fertilizers, but further experience has taught him to increase the productiveness of his soil from the resources of his own farm. This he fully coincides with us in believing to be the true principle for every farmer to act upon. It would be a lesson worth the studying for most farmers to see the economy he displays in preserving all farm manures of whatever kind. He has no fences on his farm—his horses, cows and oxen being stabled the year round. The leaves upon the woodland are raked up in autumn to serve as bedding, and it is found that they pack of their own weight so as to occupy far less room than would be supposed, while they answer the purpose admirably, as well as form a valuable constituent in the resultant manure. An apartment of moderate size serves to contain a sufficient quantity to last nearly or quite the whole twelve months. Every drop of liquid manure, from stables and styes, and brought by drains from the house and out-houses, is collected in cisterns. In it, previously to being pumped out for use, Dr. Underhill dissolves potash, in the proportion perhaps of one hundred weight to thirty hogsheads—which is thought to have the effect of making the manure more active, as well as being cheaper than ashes, in supplying the necessary ingredients abstracted by the crops from the soil. A cheap and coarse kind is bought at three and a half or four cents a pound.

"We should here devote a few words to the compost heaps we have passed here and there, in our walk over the place. These Dr. Underhill begins, say with a stratum of the alluvial deposits from the river side, followed by one of horse or cow manure or both, then one of the sods from along the roads, paths, etc., then the alluvia again, and so on. After they reach some height and when the manure cisterns chance to be full, a man perforates them here and there with a crowbar, and the liquid is brought in a cart and put on, hogshead after hogshead, till the whole is saturated. They are made amply broad enough for a cart track, extended to any length, and as they slowly settle down carried higher and higher by additional layers till six or eight feet above the ground. The same pile accumulates the manures of nine months or so, and receives four or five thorough wettings. The value of a compost heap thus prepared, in comparison with its cost, as would be readily conjectured, is very great

"In the preparation of the ground for his vineyards, Dr. Underhill thinks that thorough ploughings answer every purpose. In one case, he had had the earth trenched with spades, to the depth, we think, of three spits, but the effect produced was of too little increased benefit to pay for the expense, which was, if we recollect, in the neighborhood of four hundred and fifty dollars per acre. He adds a dressing of clay to render the soil more firm, and prevent its feeling so quickly the changes in the temperature of the atmosphere, and absorbing the rains so rapidly as to drench and chill the roots. A less quantity will answer every purpose than might be apprehended—in pretty thorough trials he had found three or four hundred loads sufficient on an acre of his rather coarse, gravelly sands. In one experiment he had spread a vineyard of about six acres with 5,000 loads of alluvia and 3,000 of clay; but it proved too rich and heavy.

"The vines are permitted to bear the first crop on a temporary trellis of stakes driven into the ground and connected by a single wire. The permanent trellis is then erected by putting in firm chestnut posts about seven feet high, and running along them a couple of wires for the second crop, and a third one near the top the subsequent season. The wire used is from number 10 to 12. We have not the space nor the necessary acquaintance with the subject to describe at length Dr. Underhill's method of pruning. It is progressive, that is, different for each of a succession of years as a vine grows older, until it finally reaches maturity. The chief object kept in view, of course modified to meet particular circumstances, is *to depend on this year's growth of wood* for next year's growth of fruit. The ground is thoroughly ploughed once in the spring, and the spaces dug between the vines; after this the harrow and cultivator are depended on to keep the soil loose and free from weeds, until the fruit begins to change its color—when no one is permitted to go between the rows until the picking begins. By thus employing horses and implements, the expense is very much less than on the German plan of merely spading and hoeing.

"When the fruit is formed in June, as much as three-fourths to four-fifths of it are cut away—only the small remaining fraction being suffered to ripen. Thus, and by a careful system of pruning, the strength of the plant is economized, and wholly devoted to the end of completely maturing the juices which form both the vine and the fruit, and adding particularly to the size and sweetness of the latter. Every effort has been made to subject the main vitality of the plant to the one purpose of producing the best fruit rather than the most wood; and by these efforts, by careful pruning, and proportioning the quantity of fruit ripened to the capabilities of the vine, Dr. Underhill estimates that he has succeeded in adding much to the strength of the plants themselves, to the excellence of the fruit they bear, and in making the period of its maturity earlier from season to season, so that there is an average difference of at least *twelve days* between the time of the ripening of the grapes now and that when he commenced his efforts 25 years ago. This appears to be no inconsiderable advantage in favor of vines from his grounds; inasmuch as we see no reason why the same causes which operate to produce this earlier period of ripening in his vineyards, should not also have a similar effect on cuttings grown from them with the same care to the age of setting out. He has plants for sale at the age of two, three, and we think also four years old, as purchasers may prefer. About one-third of the vineyards are Catawbas, the remainder Isabellas—the latter of which has been found the surest for a crop, though it is very seldom that either falls short, and we understood that for many years past, Dr. Underhill had not experienced a single entire failure. The present season has been, on the whole, a cooler one than the average—according to his expectations, based as he told us, upon the fact that *every tenth year* regularly proves a cold one—at least he knew such to have been the case for certainly sixty years back, and had no doubt it would continue so. The crop is very good, however, the berries and bunches being especially large. We saw some Isabellas that entirely exceeded in these respects anything we have seen before—here and there a cluster that must have been very nearly a pound in weight—and, although none were ripe enough to taste, we could easily credit Dr. Underhill's assurances that they would soon prove as luscious as they then looked.

"The doctor is rightly very particular that none shall go to market until they are *fully ripe.* He says it requires a good deal of experience and judgment to determine when they *are* ready for market. He expected to begin picking about the 15th, and he generally continues the marketing season for about two months. During this period he engages a store in New York, where his grapes are all disposed of, with the exception of those retailed at confectionery and other stores through the city, and consumed at the hotels. His lowest wholesale price is fifteen cents per lb.; by the basket to families, sixteen; and when less than a basket is sold, twenty. One point which we should not omit to note, inasmuch as it is one in which fruit-growers are far too generally negligent and regardless of their own interest, is the care and nicety with which Dr. Underhill prepares his fruit for market. We have seen his particularity in respect to the entire ripeness of all that are picked; and every bunch of the vast number he sells is looked over, and the defective berries cut out by hand, so that not one may be left which a child two years old might not eat with impunity. Then, put up in new and neat baskets, they present an attractive appearance, which goes a great way in winning the heart—or, perhaps, we should rather say, inciting the appetite of the purchaser.

"When Dr. Underhill commenced, a good many years ago, he was the first and only one in the business, and could only command about five cents a pound for his fruit. He has not endeavored to retain this monopoly, but is always happy to explain everything he knows to any inquirer, and justly thinks that the more good fruit he can induce others to grow, the more public attention will be brought to the luxury, or indeed the necessity of the article, and the greater will be the consequent request for it. How just were these calculations, is shown in the ample demand that now exists for all he can grow at three times the price at which he started. We trust that he will not give up his present purpose of sometime presenting to the world the system which his long and careful, and, we may add, profitable, experience has matured.

"There are several other points which we had it in mind to speak of at some length, connected with Dr. Underhill's agricultural and horticultural practice. The extent of the present paper will compel us to be very brief.

"Several lessons may be derived by every farmer from what has been already written. He has seen how our friend economizes all his manures, and how he has discovered the secret not only of constantly increasing the fertility of his lands, but of *adapting the crop grown to the wants of his nearest market*, and thereby obtaining very much greater profits than the old farm routine could in any way be made to yield. How he has proved the vineyard rules received from European authorities far from being best adapted for his situation and circumstances, and *thought out, and worked out by experiment*, a system for himself. How he has created a new demand with the public, while he was himself supplying it, and how scrupulous he always is that every product he sells shall be *superior of its kind* and put up in the *best style*. It is self-evident, we think, that not one of these particulars is immaterial to the farmer who would succeed well in his business.

"Another which we wish to bring forward, is the way in which Dr. Underhill contrives, in almost every process, if we may quote a homely proverb, to "kill two birds with one stone." Where he has dug the deposits of vegetable and alluvial matter by the water's edge, for manure, a very little extra labor has transformed the ugly excavation into a fish pond; a water gate admits the fish from the river but will not let them out, and through the same channel the rising and falling tide prevents the lakelet from lying stagnant. The pond not only supplies fish, but *plums*—the trees being planted over it at an angle of perhaps forty-five degrees to prevent the ravages of the curculio, while it is also bordered with pears and quinces, and thus the land dug out and removed is not only made to yield a crop of fruit where it is put as manure, but another over the hole it left behind. The forests are cleared out and seeded with orchard-grass, and the leaves falling in autumn are taken away for use, as we have seen, as well as that they may not smother the turf where they fell. Sods are required for the manure heap, and paths and roads tastefully, and here and there quite picturesquely threading the woods and climbing the river banks, are laid out and kept in order to yield them, as well as to furnish delightful drives and walks. It may abate somewhat from the romance of the beautiful, thus to find the useful ever lurking under its mantle, but it certainly brings it within the reach of many who now fancy it something beyond or above them, as well as places it in a new light to not a few, who are in the habit of considering themselves far too *practical* to seek it *Utile dulci* is Dr. Underhill's motto."

JUDGE CONKLIN'S VINEYARD ON LONG ISLAND.

From the Country Gentleman.

"The facts in the following description were derived from a memorandum prepared by Judge Conklin at our request, which, together with a personal examination of the vineyard, will enable us to show the actual results of his mode of cultivation.

"The first experiments were made with the foreign varieties; they grew vigorously and fruited uniformly well, especially the chasselas or sweet water, producing full crops of fine grapes for several years. After experimenting with them four or five years, they began to fail in maturing their fruit, which was supposed to be owing to the severity of the climate on this island. A subsequent trial of three or four years more proved this to be the case, after which the open field culture of all the foreign varieties was given up as useless.

"In the meantime a few cuttings of the Isabella which were planted began to produce some fruit, which appeared to be a pretty good substitute for the more delicately flavored foreign varieties, and from the date of this discovery the vineyard was commenced.

"SITUATION AND SOIL.—The grounds are located at Cold Spring Bay, around which the hills are steep and abrupt, leaving but a narrow slope of arable land between their wooded declivities and the shore; beyond these wooded hills, which rise from a quarter to half a mile, are fine cultivated table lands. Below the woods, on a western slope, lies the oldest part of the vineyard, and beyond the woods on the table land, lies the more recently planted portion. The soil, like most in this part of the island, is light and porous, composed of sandy mold and a large proportion of gravel, containing small stones from the size of a pea to three inches in diameter, mostly of polished quartz. Some spots are quite alluvial to a considerable depth, formed by the action of the rain descending from the hills; other portions approach the quality of soil called light loam, which is esteemed best for all general purposes; the under stratum is extremely porous, full of coarse gravel and small stones, with layers of sand but no clay. On account of the steepness of the declivity, some of the ground required to be terraced, the descent being so great as to wash both soil and seed into the harbor during the prevalence of heavy rains. The most abrupt portion was so barren that after it was terraced it appeared like mere banks of gravel and dead earth. In this place it required a long time to establish the vines; the rays of the sun beat down in the afternoon almost vertically during the long and severe droughts with which we were visited for several years; but after continued watering and mulching, they were at length established, and are now loaded to their utmost capacity with fruit.

"The foregoing remarks upon the soil are applicable only to that portion of the vineyard upon the side-hill; the soil of the part situated upon the table lands is of good quality for farming purposes. The table lands about this vicinity are full of gentle swells or eminences, which are more or less gravelly or porous in the substratum; such places as these have generally been selected by the judge for his vineyards, not that they were supposed to be more suited to the vine than the lower or more level parts, but because they were less capable of sustaining other vegetation.

"The first planting of much extent was made in 1830, from vines of one year's growth, purchased of Col. Alden Spooner, of Brooklyn, who furnished a thousand at six cents apiece; these were planted upon the best portion of terraced ground, which was prepared by ploughing in such a manner as to throw the furrows down hill, and then finished by hand labor. A part of the terraces were made ten feet wide, but as the operation was found laborious, the remaining ones were made only eight feet; no particular pains were taken in forming banks.

"The original surface being a sward, the sods were placed on the outside of the terraces to sustain the banks. They soon covered them with gra s, which has since needed no further attention excepting to mow it once or twice a year. The vines were planted six feet apart, with a locust post between each, and then four lines of No. 12 wire were strained and fastened firmly upon the posts, beginning two feet from the ground and setting the wires one foot apart; the posts, therefore, require to be six feet high, which is the usual height throughout the vineyard.

"The vines have not been subjected to the rigid system of pruning practised by many; they were trained somewhat fan-shaped upon the trellis and rather slovenly pruned for many years. The plough was freely used, seldom allowing the ground to get hard and weedy. This vineyard bore uniformly for many years, yielding great crops of grapes, which commanded a fair price in market.

"In a few instances of late, it has been prostrated for a year or two from the effects of enormous bearing. This was the case last year, but it is now loaded with fruit. The vines which were placed upon the widest terraces have been much the most productive, showing that it would have paid better if they had all been of one width.

"A few of the vines purchased of Col. Spooner were planted on the more level part of the ground; these also succeeded well, but not so uniformly as those placed on the terraces. About the same time, one hundred three year old vines were purchased at the Parmentier garden in Brooklyn at 37½ cents per vine; these were also put upon the lower grounds, and were six feet apart each way, planted very deep and the earth gradually filled in around them. The subsoil was almost entirely composed of small stones and gravel without a particle of clay, or even loam, to be seen. They grew rather slow at first, but soon got established, and are yet bearing full crops, failing however oftener than most of the vines planted at a greater distance apart. The next portion were planted near these rows running north and south, eight feet apart and six feet apart in the rows; these were set out at an ordinary depth in the ground, and treated in the same manner as the foregoing, and the vines have yielded good fruit more uniformly than any of the preceding ones. Another piece was planted in the midst of winter, during a season of mild weather, while the frost was out of the ground; hardly a vine failed, and they have borne largely with few exceptions.

"The last piece, comprising about three acres, was planted on a good strong soil, not heavy, but sufficiently loamy for most purposes; which was previously occupied by locust timber grown from seed that had been so much injured by the grub that it was deemed good policy to substitute a vineyard.

"The vines promised much, and have borne pretty well, but have fallen short of anticipations, owing perhaps to peculiarities of the season, which have been marked for the last few years. They were set in rows running north and south ten feet apart, and the vines eight feet apart in the rows.

"No difficulty was found in subduing the locusts; a contract was made with a laborer to dig each tree out, removing the soil and extracting the entire root from a circle of the diameter of a cart wheel. This work was performed in the winter at a shilling a tree; the ground was then ploughed and planted with corn; the following year a line was designated for each row of grape vines; the ground was then ploughed to the width of four or five feet along these rows, throwing the furrows out until a considerable trench was found, while workmen followed with suitable tools and cut away all interfering roots. This operation gave space for the row of grape vines on clear ground, with but little necessity for removing much earth by hand for their reception; the remaining roots in the middle of the rows were soon got rid of by the subsequent ploughing after planting the vines.

"Failures have occurred occasionally, owing to several causes, some of which might have been avoided; among these may be mentioned the rose bugs; these came in such quantities, after several years, as to entirely destroy the crops for one year, before their existence was hardly suspected. After this they were caught by carrying small pans of water along the rows, and shaking them into it; a gentle agitation being sufficient. For two or three years they were very troublesome, and required constant looking out for; they were got under, however, and little trouble has since arisen from that quarter; a few yet linger around and make their appearance each year, but a sharp lookout is kept, and they usually do little damage.

"Excessive bearing is a common cause of failure, perhaps the most general; it is one of the peculiarities of the vine to set enormously with fruit when strong and vigorous; the tax upon its maturing such a load often produces complete prostration; sometimes it will show its effects the first year—indeed this is often observable in the sour, shrivelled mass of unripened fruit.

"It will frequently happen that the vines will go on and mature the present crop, but they are sometimes so completely prostrated that they require two years to get restored. Pruning thoroughly in the winter, rubbing off all superabundant shoots, and cutting off a large portion of the clusters of fruit in the summer, are the proper remedies.

"Excessive rains during the months of May and June have been more destructive than any other cause, especially when accompanied with unusually cold weather about the time of blossoming and setting. Two years nearly the whole vineyard failed from this cause; all over the most retentive and best portions of ground, the young shoots that were about putting forth blossoms, turned black and sour within a few days after one of these heavy rains, and the crop, which until that time promised to be a fair one, was ruined, leaving a small yield upon the highest knolls.

"Another cause of partial failure is, perhaps, not so obvious, yet not less sure in its effects; we allude to certain peculiarities in the atmosphere during some seasons, which seem to be very uncongenial to the growth and maturing of our best fruits; probably these effects are felt more or less all along the Atlantic coast. They have forced us to abandon the culture of the old Virgalieu Pear, and the White Chasselas, and other hardy foreign grapes, while all began to feel its influence here at the same time, and it has nearly vetoed the Newtown Pippin, and many other apple-trees are exhibiting like effects in a less degree. The indications are the same upon both pears and apples; they exhibit a rusty coat, cracked open, and are hard and bitter where these black spots exist. On the Newtown Pippin they are more obvious on the northeast side of the tree, the fruit often being quite fair on the southwest side, while on the opposite side it is nearly worthless.

"The judge states that whenever a Newtown Pippin bears a crop of good fruit—which it does occasionally—his vineyard yields a large crop, such being the case this year. The trees are growing in various places among the vines, and are full of fine-looking apples—everything in the shape of an apple seems to be fully developing, and so of the vines. And *vice versâ*, whenever the fruit fails of wholly maturing on these trees, there is a very marked, tardy and imperfect development of the clusters of grapes; latterly these coincidences have been very decided. The judge has resorted to training on arbors, which, he thinks, if properly managed, is by far the most certain mode of obtaining fine fruit, as there is much protection effected from the cold winds by the overshadowing leaves, for the fruit always hangs underneath.

"He thinks eight feet not sufficiently far apart for the vines, and has removed several hundred, leaving a space of sixteen feet between the vines, with a design to train horizontally along the trellis. This mode fully answers his expectations; it gives fine fruit, and it will fully equal the others in quantity after one year.

There is scarcely any limit to the capabilities of a vine, the roots always keeping pace with the top. When they are dwarfed by constant pruning, the roots are circumscribed in proportion, and consequently draw their supplies from comparatively small space; in severe drought such vines feel its effects very sensibly. The judge says this idea was very much impressed upon him in observing the roots of the vines removed in his vineyard, while at the same time some large bearing vines of the same age, growing on trees where full scope was allowed, had roots of twice the size and length.

"The stock of young vines for planting for the last 20 years has been raised from cuttings taken from the vineyard.

"No. 12 Pennsylvania wire is used; and we noticed that instead oi being fastened on to the posts with staples or nails in the usual manner, a cut is made with a small saw, and a turn taken around the post with the wire, drawing it up as snugly as possible into the cut; the wire requires forcing in a little with the hammer.

"The first part of the vineyard has had no new wire yet, except when from accident the old was broken and required repairing. It looks as though it might last ten years longer. Smaller wires have been used, but they do not answer as well. If any change was made, it should be for the size larger.

"The judge uses the common manures collected about the farm-yard; applying them in moderate quantities as best suits his convenience—chips from the wood-pile, and even shingle shavings, have been applied in large quantities during the prevalence of severe drought, with beneficial results, which have extended to subsequent seasons.

"He is now cultivating the Catawba, but not very extensively, and does not consider it quite so certain in maturing, and finds it more difficult to propagate.

"Eight or ten more native varieties are now under trial."

VINEYARD OF C. McKAY, NEAR CANANDAIGUA, N. Y.

From the Horticulturist.

"Some time ago, you may remember, you invited me to communicate to you such facts for publication as I might have met with in grape culture that would be likely to be of interest to the public.

"I had then recently planted one acre of *Isabella* grape vines, pretty nearly after the manner you had advised in the columns of the 'Genesee Farmer.'

"The piece of ground planted is twenty rods in length by eight in width, and was planted five years ago last spring, in the following manner: About the first of May I gave the land, which is gravelly loam, a very deep ploughing—as deep as possible without the aid of a subsoil plough. I then measured it off into eight strips, or lands running lengthwise, their direction being from north to south, 15 degrees east, and ploughed these lands separately—leaving the dead furrow in the centre of each, designating the places for the rows—breaking up the yellow subsoil by repeated ploughing, through the centre of each to the depth of nearly two feet. I then went into these trenches with a stout team and scraper and excavated holes a rod apart still deeper than I had ploughed, about six feet wide and eight in length, leaving the subsoil taken from them in the intervening spaces.

"All this time I had my eye upon a *drove of cattle* (some eighty head), which had died in this town the previous March and April, while performing a pilgrimage from the far West to the New York market. These I procured of the proprietor, and had them cut into pieces of convenient size, and hauled to the field and placed in the holes prepared for their reception. There being one hundred and sixty holes, a half of a carcass was placed in each. This being done, the holes were filled about half full of good surface soil; upon this I distributed as equally as possible among all the holes, sixteen heavy loads of decayed leather shavings, from a currier's shop, the accumulation, as I was informed, of about twenty years. A sufficient quantity of surface soil was thrown upon these, and thoroughly incorporated with them, to fill the holes rather more than level with the surface of the ground. Now about a bushel of well rotted stable manure, taken from under a stable, well mixed with about the same quantity of charcoal dust, from an old coal pit, was spaded into each place designated for the reception of a vine.

"I then procured of Elwanger and Barry, good strong two year old vines, with which I planted one half of this ground; and the other half I planted with layers of the previous year's growth, without a particle of top to any of them—each consisting simply of a short section of a vine of the previous year's growth, with one bud and a few small roots attached to it.

"These vines have had no other manuring since they were thus planted, excepting about two bushels of leached ashes forked in around each vine last season, and about one quart of plaster applied to each the season before. They are trained on trellises running from north to south, eight feet high, made of chestnut posts (for want of cedar), five inches square at the bottom, and two and a half by five inches at the top, set eight and a quarter feet apart, with strips of one and a half inch stuff, two and a half inches wide, nailed from post to post, eighteen inches above the ground, and at the top of the posts. Between these, three tiers of No. 14 iron wire are drawn, dividing the space equally between the wooden strips, and secured to each post.

"These trellises are now completely filled with good, strong, bearing wood ready for use next season, much of which is over three quarters of an inch in diameter, and large portions of it are now apparently ripe. I allowed these vines to bear only about seven pounds each, last season; though they were set for full three times that quantity. I rubbed off every alternate bud on all the vines last season; and after they were set for fruit, I took off half of it. My fruit was mainly sold to dealers in Elmira, and retailed by them at fourteen cents per pound, by the side of *Isabella* grapes, cultivated near Penn Yan, at twelve and a half cents. One dealer, Mr. H. H. Richards, afterward informed me that he sold fifty-three pounds of my grapes in one evening at fourteen cents, and but three pounds of the shilling grapes. Do you suppose those dead carcasses had anything to do with this? I do.

"Last spring, before these vines commenced their growth, I measured some twenty-five or thirty of them, taking them 'as they run,' and I found but very few of them to measure only ten inches in circumference. Nearly all measured over a foot around the body, several of them fifteen inches, and one seventeen inches. But why did not those dead cattle and leather shavings kill them? Surely it is a marvel that they did not; for I have repeatedly dug down to the bones within the past two years, and have always found them completely surrounded with a net-work of *living* fibrous grape roots—not *dead* ones! I am allowing these vines to bear this season just half of what they set for, after a severe autumnal pruning; and I estimate the present crop at 3,200 pounds, or 20 pounds to the

13*

vine, notwithstanding the hail storm on the 4th of July destroyed at least 1,000 pounds.

"My grapes last season commenced making their first turn on the 1st day of September, and the entire crop was ripe before the 30th. This season they commenced turning red on the 20th of August, and at this time (September 7th) more than 1,000 pounds are making the second turn. In fact, I have seen *Isabella* grapes offered for sale in Rochester, in the month of October, not as ripe as these are.

"I will not say *positively* that my fruit is equal in flavor to that produced by more seemly fertilizing materials, for that would hardly seem possible; but I will send you a sample of it as soon as I consider it ripe, that you may have an opportunity to judge for yourself. One thing I have remarked in regard to these vines; no insects of any sort have disturbed them, except that three or four of the vines, the year they were planted, were dug out by dogs in their nocturnal attempts at a premature resurrection of those dead carcasses! Mildew has never affected them, although, from their remarkable luxuriance of growth, I have expected to encounter much trouble from this most patience-trying scourge of the vineyard.

"I have adopted the plan this season of mulching my vines with spent tan. I applied it early in July, having kept the vines thoroughly cultivated previously; I shall cover the entire vineyard with it next season about one inch deep. I have always ploughed between the rows to the depth of five or six inches, to within a foot of the vines. To enable me to do this without disturbing the roots, I have practised pruning off all the lateral surface roots, to the depth of five or six inches, thus throwing the vines, to use a familiar phrase, upon their 'taps.' To this practice, I believe, ought to be attributed any exemption from mildew, far more than to any or all things else. I observed this same practice prevailed in Dr. Underhill's vineyards at Croton Point, though I am not aware that the doctor has ever given the fact to the public. If he has not, of course it is because he forgot to do so; or he may have thought it would not be of much interest to the other cultivators of the grape. One fact is worthy of note: Dr. Underhill has experienced no difficulty from mildew for quite a number of years past, and his vines are very old; whereas when his vines were young, he says he was much troubled with its presence. Now, all who know anything about grape culture, well know that old vines, with ordinary culture, are far more subject to this difficulty than young ones. To my mind, this proves that the doctor is older than he once was, as well as his vines, and that he has not grown old to no purpose.

"In conclusion of this already too long article let me say: if you do not like the samples of grapes sent, suppress this account of their origin and culture, for they are the *argument* I must rely upon in defence of my mode of grape culture. If this argument fails to produce conviction, I will yield the point; but if you like them you may give me a hearing in the columns of the 'Horticulturist,' if you choose to do so, that others may learn by what strange means good fruit may be produced, in spite of the abuse so profusely heaped upon 'dead carcasses' by those who never take the trouble to give them a patient trial."

APPENDIX II.

CULTIVATION OF THE VINE UPON TRELLISES (EN TREILLE) IN NORTHERN AND CENTRAL FRANCE, ACCORDING TO THE NEW METHODS IN USE AT THOMERY.*

THE table grape cultivated in the open air acquires often in Central, and with greater reason in Northern France, only an imperfect maturity and mediocre quality, for want of proper and sufficiently prolonged heat during the summer. The vine starts with vigor, but its growth is too much prolonged, and the ripening is not completed by the first cold weather of the autumn; for it is only when the sap channels cease to feed the clusters that the grape begins to ripen. This prolonged vegetation is also the reason why the shoots are but imperfectly formed, or matured by the August heat, and why the vintage of the next year is less abundant. To avoid this cause of failure, the vine is disposed in the form of a *trellis*, upon walls placed so as to enjoy the best exposure, and soils are chosen of a light or medium nature, which are easily drained and warmed; lastly a series of operations is applied to the vine, the result of which is to maintain it in a state of moderate vigor, and above all to diminish the period of its yearly vegetation. The trellis of the Château of Fontainebleau was the first which, in its culture, taken as a whole, best fulfilled the

* The first trellises at Thomery were established about 120 years ago by a cultivator named Charmeux, grandfather of the present M. Baptiste Rose Charmeux. He built the first wall for the purpose, leaving in the centre, according to a condition imposed upon him, a gate for the passage of the chase.

conditions which we have just indicated, and it has been chosen for a model by all the authors who have written upon the cultivation of the vine *en espalier*. This trellis, 1,500 yards in length, was put up nearly a century ago, and was restored about the year 1809 under the direction of Monsieur Lelieur. But long before the last named period, the inhabitants of Thomery, a village five miles distant from Fontainebleau, were adopting entirely this method of culture. They found in it so much advantage that they finished by covering with walls intended for the vine the greatest part of the territory of the Commune.

This culture at the present time extends over more than 3,200 acres, and produces on an average a million kilogrammes of grapes. It is the delicious produce of these trellises which are sold at Paris under the name of *Chasselas de Fontainebleau*, Fig. 68. Encouraged by their success, these intelligent husband-

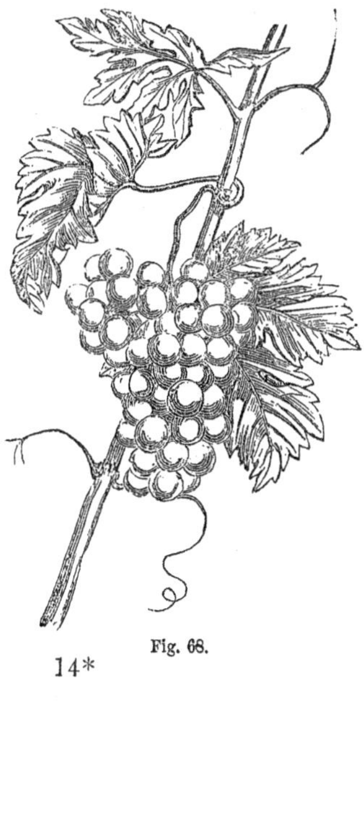

Fig. 68.

men have continued to perfect their processes, and the greater part of their trellises are at the present time arranged and maintained much better than those of Fontainebleau. The reader, however, would be in error should he believe that the success of this method at Thomery is due to the soil, to the climate, or to the exposure of this locality being particularly suitable to the vine. The soil through most of the commune is of a clayey nature, and retains a slight dampness unfavorable to the quality of the grape. The ground is generally inclined to the north-east, and, lastly, the neighborhood of the forest, by which the commune is surrounded on one side, and that of the Seine, by which it is bounded on the other, maintain a humid atmosphere very injurious to the vine.

It is chiefly to the skill of the cultivators that we must attribute such happy results. We shall, therefore, describe the mode of culture practised by them, and recommend it for the climate of the centre and the north of France.

FORM TO BE GIVEN TO THE TRELLISES.—The form the most commonly adopted until quite lately has been that of a simple

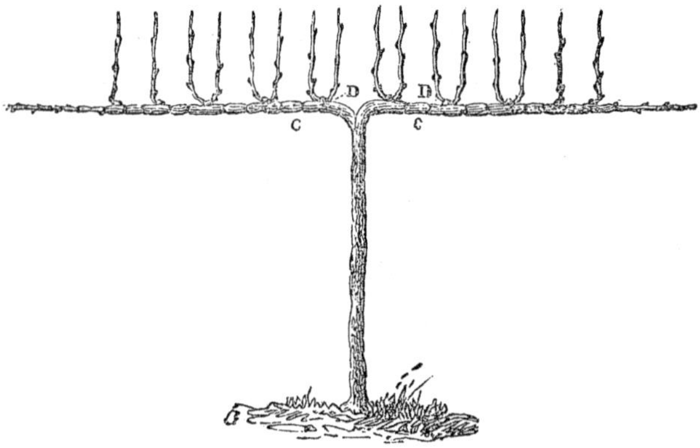

Fig. 69.

horizontal branch (*en cordon horizontal simple*), Fig. 69. It is the best form for allowing the action of the sap to spread equally toward all points of the plant, and at the same time it occupies without loss of space all the surface of the wall. But these cordons, *or arms*, must be subjected to certain conditions.

First. The two arms should present exactly the same length, or else it will be seen that the longer arm will absorb the greater part of the sap and soon destroy the shorter. Moreover, the shoots which these arms bear should spring only from the upper surface and at regular intervals of from seven to eight inches.

Second. The entire length of the arms developed by the same stock should not pass certain limits, for if they are permitted, as is often the case, to attain a length of from 32 to 48 feet, the sap tends principally toward the extremities, the shoots growing upon these points are too vigorous, while those nearer the origin of the arms become feeble and finally wither. It is much more profitable to increase the number of stocks against the walls, and to concentrate the action of the sap in a less extent of branches. In light soils and to varieties of ordinary strength, an average length of 52 inches is given to each one of the arms (or cordons) of the same stock. This, in very fertile soils, may be increased to 66 inches. With respect to very hardy varieties, as the Frankenthal, a length of from 78 to 97 inches is allowed. That adopted at Thomery is commonly 93 inches.

Third. The same stock should not bear many cordons one above the other, for the sap tending principally to the upper cordons, those beneath will remain weak.

Fourth. In many gardens may yet be seen vines fixed to the upper part of walls against which are trained different sorts of fruit trees. This is a very bad arrangement. If the cordon is placed in the most favorable condition for ripening the grape, that is 19 inches lower than the coping of the wall, the foliage of the vine shadows the trees trained below and condemns from 11 to 15 inches of their tops to complete sterility. Moreover, they deprive these trees of the influence of the rains and dews of summer. If, in order to avoid these inconveniences, the cordon is placed above the coping of the wall, it is only with great difficulty that the clusters, no longer protected, arrive at maturity. It is better, then, entirely to abandon this arrangement, to devote a certain space of wall to the vine, and to proceed in such a manner that this shall be entirely covered. This is what has been done for the trellis of Fontainebleau, and those of Thomery, by means of the following forms.

Horizontal Cordon of Thomery. Fig. 70.—Each vine-stock taken by itself presents exactly the arrangement of the simple horizontal cordon. That which constitutes the Thomery system is the position of the cordons with regard to each other. The wall is covered from summit to base with cordons of the same length placed one over the other, and supplied by vine stocks planted at regular distances.

To construct this trellis we first determine the distance to be preserved between each *cordon*. As the space is to be filled by shoots which spring from the upper surface of the cordons, it should be such that the shoots may reach a development sufficient to maintain the requisite degree of strength in the vine, without, however, passing the upper cordon, for it would, in that case, be shaded too much. Experience has shown that a distance of from 17 to 20 inches is, in the greater number of cases, sufficient, and that to this height the shoots may be deprived of their buds without diminishing the strength of the vine. This distance, may, however, be augmented for very hardy varieties in very fertile soils by from 4 to 6 inches. Monsieur Felix Malot has established at Montreuil a trellis, the cordons of which being placed at a distance of only 15 inches from each other, render it necessary to stop the shoots as soon as they have attained that length. The sap from the roots being concentrated in a smaller space, he obtains, in general, larger bunches; but this detracts from the strength of the vine and the duration of the trellis, and the growth of the grapes occupying a longer period, they do not ripen so well. The cultivators of Thomery prefer smaller bunches more equally ripened. It will next be proper to decide the height of the wall, that we may know the number of cordons to be erected. Supposing that, like almost all those of Thomery, this wall is $8\frac{1}{2}$ feet in

Fig. 70.

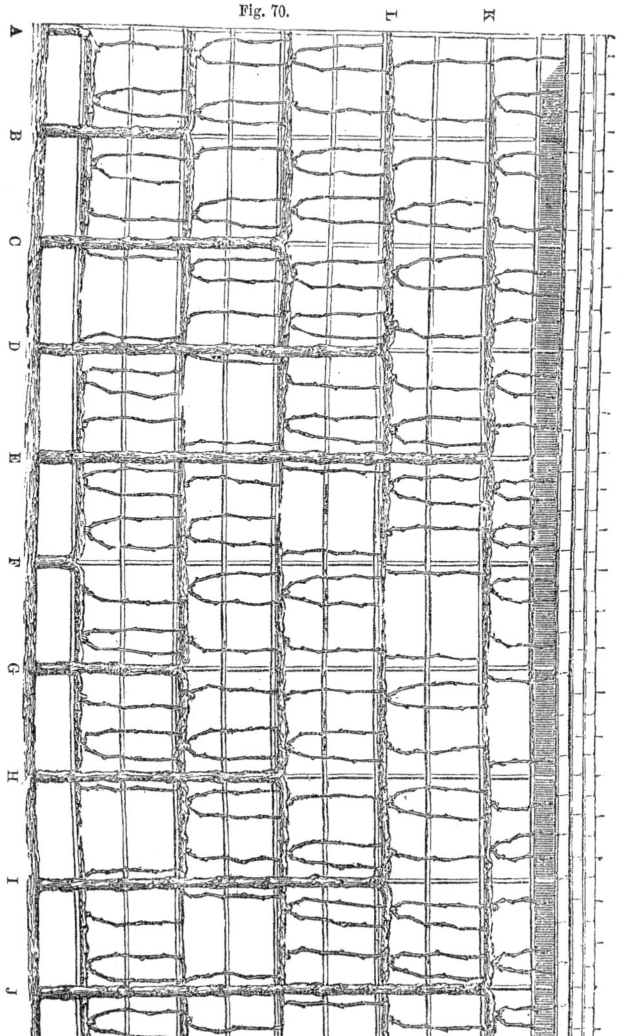

height,* by dividing this number by 17 inches (the distance of the cordons from each other), we obtain just six inches. The first cordon being established at 15 inches from the ground, we shall then be able to place upon our wall five cordons.

As to the distance to be preserved between the stocks, that is necessarily determined—first, by the length to which the two arms are to be allowed to grow; and secondly, by the number of cordons one above the other. Let us suppose this number to be five, and the total length of the two arms to be 8 ft. 10 in. To know the distance sought, divide the total length by the number of cordons: we obtain 20.12 inches, which we have adopted for our figure. At Thomery, where the cordons have only a length of 88.8 inches; the stocks are planted at intervals of seventeen inches. It might happen that the wall for the trellis might be less than 8½ feet in height, and that the number of cordons being reduced from five to three, the distance between the stocks will then be 33.99 inches. But an interval so great exposes the trellis to a degree of growth prejudicial to the ripening of the grape; in that case it would be better to diminish the length of the cordons from 8⅔ to 5.87 feet, and the distance between the stocks will then be 19.76 inches. It may also be that the wall will be more than 8½ feet in height, and in order to increase the number of cordons it becomes necessary to place the stock at less than 20.12 inches; for example, to 9.88 inches if the wall affords space for ten cordons. Should this distance be too small to allow the roots to draw from the earth the sustenance necessary for the support of ten cordons, in order to remedy this inconvenience the length of the cordons is slightly increased for a trellis of ten cordons to from 103.08 inches to 62 inches, the stocks remaining at the distance of 15 inches from each other. However, as this increased length of the cordons has an unfavorable

* M. Du Breuil has had his drawings made to an accurate scale, but although our figures are exact transfers from his cuts, the relations of the French and English measures is such as to preclude our giving a useful scale in English feet. The same reason has obliged us to introduce numerous fractions—a feature which we did not feel at liberty to avoid by violating the accuracy of the translation.

influence over the vigor of the branches and the quality of their products, we recommend in preference that the following process be employed.

It consists in planting on the side of the wall which is to

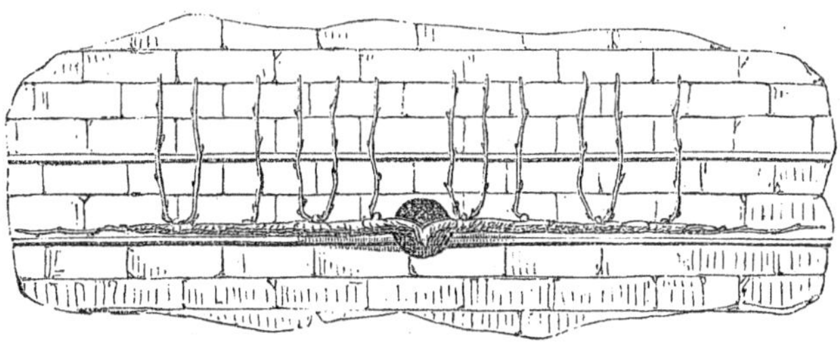

Fig. 71.

receive the trellis only the number of stocks sufficient to form five cordons, at the most. As to the other five cordons, if the height of the wall requires ten, they should be established by means of stocks planted *on the other side* of the wall, and which may pass to the front through holes pierced in the wall at each point where a cordon is desired (Fig. 71). When the vines have grown through the wall, the opening should be closed with clay in order to avoid injurious currents of air. The cordons formed in this manner are preferred by the cultivators of Thomery for the lower cordons of the trellis. They have remarked that the vines planted on the shady side of the wall present a stronger growth than the others, doubtless because the soil is less dried by the heat of the sun, and that the greater part of their stems escapes the action of the solar rays. Should these vines form the upper cordons, the size and abundance of their leaves would injure those below. In placing them, on the contrary, on the lower part of the trellis, their too great growth is diminished, and their clusters nearer the ground are subject to a more elevated temperature, by which their ripening is hastened.

This ingenious method may also be used for trellises composed of five cordons, but which are placed in a soil so dry and scorched that the distance of 20 inches between the stocks is not sufficient to allow the roots to gather from the earth the nourishment required. This distance must then be increased, without however augmenting the length of the cordons.

When the position to be given to the cordons is properly determined, the plan of their arrangement is traced upon the wall. Begin by indicating at the foot of the wall from A to J (Fig. 70) the point from which each shoot should spring, and from that point draw a vertical line.

At the point A, this vertical line ceases at the height of the first cordon at 15 inches from the soil; at the point B, 33.08 inches; at the point C, at 50 inches, and as far as the point E, where the line of the first cordon ceases at 86 inches from the soil. From thence a second series of lines is commenced similar to the first, and we proceed in this manner to the end of the wall. It then only remains to trace, touching the top of each vertical line, the course to be taken by the cordons from right to left, and to indicate where each of them is to cease—that is to say, at 52 inches from each side of the main stalk. After the conclusion of this operation, the vines are planted in the manner which we intend to describe.

HORIZONTAL CORDON OF CHARMEUX (Fig. 72).—The arrangement which we have just explained is that which was at first generally adopted for the Thomery vines, and is that which is still employed for the trellis of Fontainebleau. But it was not long before the cultivators of Thomery remarked that this method presented an important inconvenience. During the formation of the cordons, an entire arm of each stock is shaded by the upper cordon, while the greater part of the opposite arm escapes this unfavorable influence. The result is an inequality of growth between these two arms, and it becomes necessary to employ certain processes, often unavailing, to maintain a proper equilibrium of growth between the two arms from the main stalk. To obviate this difficulty, in 1828, M. Charmeux, senior, invented a new description of horizontal cordon, which has been adopted by almost all the cultivators of Thomery for the trellises which they have since erected. The plan is as follows:

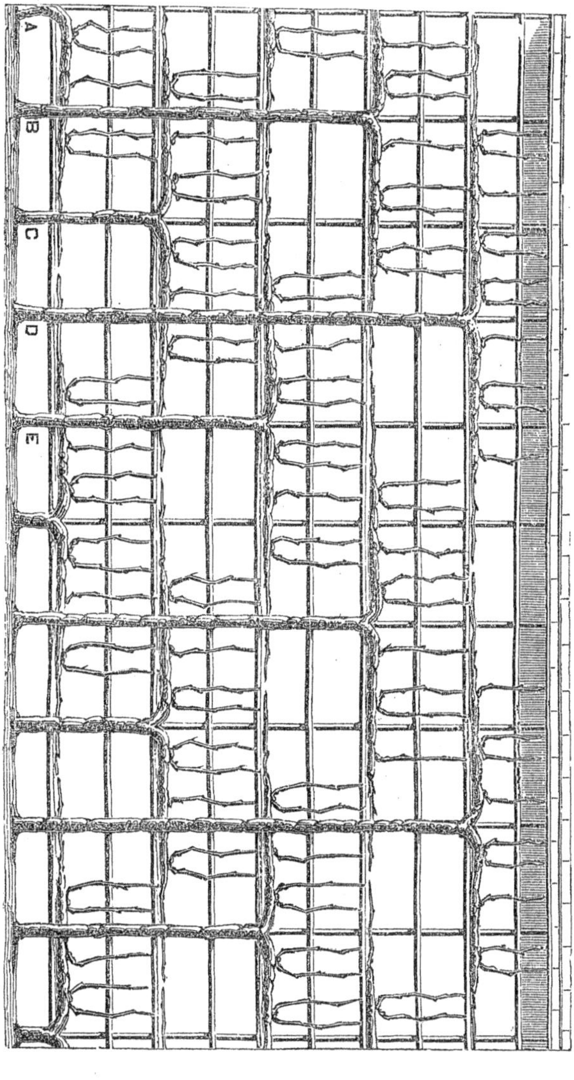

Fig. 72.

The distance between the superposed cordons, the length of the latter, and the distance between the stocks, is the same as in that of the horizontal cordon of Thomery. The Charmeux cordon differs only in the order in which the stalks successively put forth the cordons forming the trellis. Thus, in the Thomery cordon, the first stalk (A, Fig. 70), produces the first lower cordon; the second stalk (B) the second cordon, and thus to the highest cordon of all, in such a fashion that the whole number of stalks form, from one extremity of the trellis to the other, a succession of distinct steps. On the contrary, in the cordon of Charmeux, (Fig. 72), the first stalk (A) furnishes the first cordon, the *second* (B) the *fourth*, the *third* (C) the *second*, the *fifth* (E) the *third*, to commence again by the first cordon, and continue in the same manner to the end of the trellis.

The design of this trellis upon the wall is made as easily as for the preceding arrangement.

The desired purpose of this contrivance is completely attained. Not only the cordons are not unequally shaded, during the first years of their growth, but they completely escape this influence until the age of about five years. If then they are subjected to this shade, it is equal for both arms, and is brought first to bear upon the ends of each cordon in such a manner as to moderate their growth to the advantage of the bearing shoots nearest to the main stalk.

Vertical Cordon.—This disposition, to which, absurdly enough, the name of "palmette" has been given, has been applied to the trellises of Fontainebleau to a small extent for about forty years, and ten years later to some trellises at Thomery. The following is the principle. The vines, planted 39.37 inches apart, are allowed to develop a single stalk, which rises vertically to the top of the wall. This stalk presents on each side a series of branches irregularly disposed. The shoots annually developed from these are trained obliquely in the space by which each main stalk is separated.

This system of is susceptible many improvements. It is evident, for example, that the interval of one metre (39⅓ inches) which separates each main stalk, is too great when the shoots are trained obliquely and not perpendicularly from the main stalk upon which they grow, as is the case in the trellis of which we have spoken. Moreover, the irregularity with which

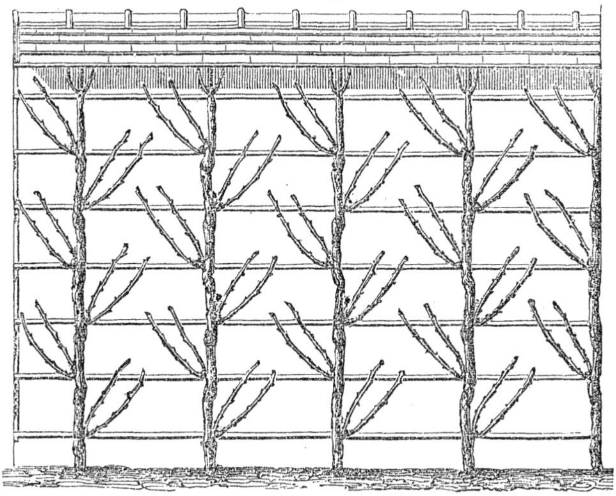

Fig. 73.

the branches are distributed upon the stalk causes an unequal distribution of sap, and its determination to certain points, whether of superabundance or scarcity, resulting in the destruction of the shoots less favorably situated.

Vertical Cordon with alternate Shoots (Fig. 73).—M. Rose Charmeux has brought this new arrangement to perfection in the following manner. He plants the stocks at a distance of 28 inches one from the other; then he regularly distributes the shoots on each side of the stem, making them spring alternately every 10 inches in such a manner that they may be separated by an interval of 20 inches on the same side of the stem. We shall find, in discussing the method of pruning, how perfect regularity in this respect may be obtained.

The trellis thus arranged presents the following advantages. In scorched and dry soils, the stocks and the horizontal branches arranged in the manner previously described suffer much from the heat of the sun, from which they are very imperfectly shaded by their leaves. In the trellis with alternate shoots the main stalks are completely covered. These cordons may therefore be usefully employed in dry soils. Besides, these cordons are suitable for the most confined space, since they require only 28 inches.

But this vertical cordon cannot be conveniently applied against a high wall, for as the sap tends toward the top of the plant, the shoots toward its base become feeble and languishing. We have remarked this fact at Fontainebleau, where the wall which supports these cordons is 13 feet high. It is our opinion that the main stem should not be allowed to exceed 80 inches. If the wall is higher, the following modification (Fig. 74) may be used, equally due to M. Rose Charmeux. For a wall 13 feet high the stocks are planted only every 14 inches; then the stalk of each is allowed to rise alternately to 66.4 inches and to 13 feet; but the latter commence to bear shoots only directly above the point where the first cease—that is to say at 66.4. In this manner the wall is completely covered and the cultivator has not to dread the destruction of the lower shoots.

The trellis with vertical cordons which we have just described is simpler and more easily formed than those with horizontal cordons; but experience has shown that its produce is less abundant, since for an equal surface it offers a smaller number of branches.

M. Rose Charmeux, struck by the advantages offered by the simplicity of this arrangement, has attempted to render it as fruitful as the horizontal cordons. He completely resolved the problem in 1828 by means of the following modification, which gives for the same surface of wall a greater number of shoots, and consequently a greater number of clusters. As this new arrangement is at once more simple, and more easily obtained than the others, and as it may be accommodated to walls of all heights, we recommend it to the exclusion of other plans, and

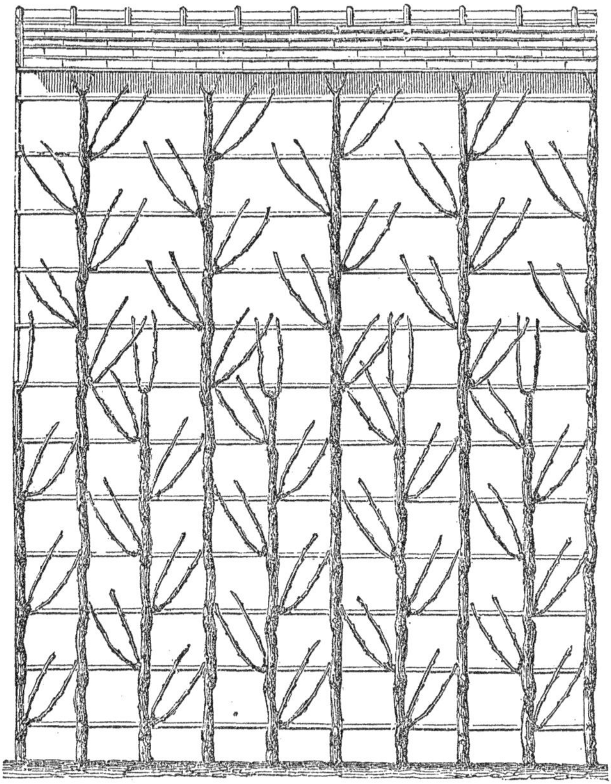

Fig. 74.

we shall choose it to study in detail the method of cultivation and pruning suitable to trellised vines.

CULTIVATION OF TRELLISED VINES, ARRANGED WITH VERTICAL STEMS (CORDONS) BEARING OPPOSITE SHOOTS.

In this new arrangement (Fig. 75) the vines are planted at the foot of the wall every 14 inches. The wall, whatever be its height, is horizontally divided into two equal parts. The first vine stops at half the height of the wall. The second is allowed to reach its summit, and thus continue in this manner alternately to the extremity of the wall. The reader will observe that the shorter vines bear shoots from about 12 inches above the soil to their tops, and the taller begin to bear shoots only on leaving the lower half of the wall. These pairs of shoots are 10 inches distant from each other. This arrangement offers all the advantages presented by the form shown in Fig. 74; that is to say, that in consequence of the length of the main stem furnished with shoots, these last are maintained in equal growth. Moreover, the new form grows more shoots for the same surface than is shown in Fig. 74, and more even than the horizontal cordons. If, however, the wall is only 39 inches high, all the vines may be made to rise regularly to its summit. But in that case they should be placed at intervals of 28 inches and should be furnished with branches from 12 inches above the soil to the top of the wall.

Let us now turn to the labor necessary to the establishment of such a trellis as the one described above.

Walls proper for the Trellis.—The vine arranged in vertical cordons accommodate themselves to walls of all heights. At Thomery the gardens are subdivided by bearing walls parallel to each other and separated by a space of from 40 to 46½ feet. They may, however, be placed nearer to each other, but in that case the earth between will be too much shaded, and cannot be turned to account These bearing walls are only 85 inches high, and were built many years after those of inclosure; that is to say, when the young vines which it is intended they shall support, have been carried thence by many successive

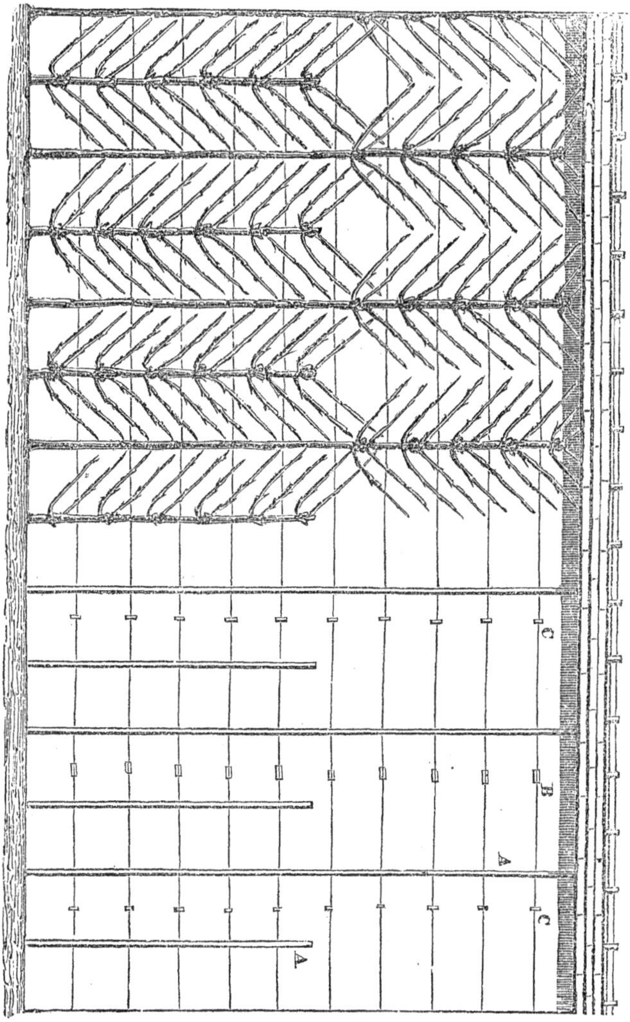

Fig. 75.

layerings (*couchages*). Thus the interest of the capital employed in these constructions is economized. Some of the cultivators of Thomery have also constructed a sort of *counter espalier*—that is to say, a lesser wall opposite the principal bearing wall, in masonry 45 inches high and 6 to 8 inches thick. Only one of these little walls is placed 100 inches in front of the principal walls the most favorably situated. In this manner they derive every possible advantage from their best exposures.

This subdivision of the inclosures not only enables the cultivator to obtain a larger harvest, but it likewise offers the advantage of diminishing the currents of air, concentrating the heat by the radiation, and thus hastening the ripening of the grape.

It has sometimes been attempted to use for trellises the walls by which terraces are supported. The superfluous moisture of the soil draws to the bottom of the wall and injures the vine stalks. For almost every other kind of fruit-tree very projecting copings offer more inconveniences than advantages, but for the vine the case is different. On the one hand, these copings take place with those movable fruit-houses which we have recommended for covering espalier trees in order to preserve them from the chills of the spring; and on the other, they shelter the vine from the moisture of the rains and dews, which results in a more active vegetation and a more prolonged development, injurious to the ripening of the grape. Finally, these projections preserve the clusters from the first cold weather of the autumn, and thus delay the time of gathering and facilitate their preservation. All the walls of Thomery are finished by tile copings. Their projection is greater in proportion to the height of the walls, being 14 inches for walls of 156 inches, 12 inches for those of 117 inches, 10 inches for those of 100 inches, 8 inches for those of 80 inches, and 5½ inches for the little walls of the *counter espalier*. In the last case they are inclined only from one side.

The walls thus built are white, being covered with lime. This color at Thomery has given the most satisfactory results.

When the method of construction allows, smooth finished walls (*palissage à la loque*), should be used; we may then dispense with the trellis. But the great quantity of plaster required by this arrangement renders it too expensive to be used beyond a certain distance from Paris. We must have recourse to frames, and for the form of trellis of which we are speaking, they should be erected in the following manner:

A series of galvanized iron wires (No. 14) are extended along the wall. Upon these wires laths are fastened every 12 inches, and to these laths the main stem of each vine is trained alternately to half the height and to the summit of the wall.

Exposure of the Walls.—The trellised vine demands an exposure at once as dry and as warm as possible. In the north and the centre of France this double condition is best fulfilled by a southeast exposure. A southern exposure is doubtless the warmest, but the trellises with such an aspect also receive too directly the damp winds and rains of the southwest. The cultivators of Thomery use the side of their walls exposed to the west and to the southwest, but gather from thence grapes of the second or third quality only.

Propagation of the Vine.—On the different modes of propagation to be chosen for a trellised vine, we would offer the following observations. Slips or cuttings propagated from layers are often used in forming trellises. When intended for a permanency, they are planted in the manner which we will proceed to describe. They begin to bear fruit only in the fourth year. They should be used only in the absence of the layers themselves, for whose first fruit we are not obliged so long to wait. The layers, or as they are called at Thomery, the *chevelées*, are generally to be preferred, for when they are transplanted with care, and their roots are not dried by exposure to the air, their vegetation during the first years is more vigorous, and thus time is gained. Two sorts of layers are used—uncovered layers and layers in baskets. The uncovered layers (Fig. 76) are freed from all the earth which surrounds them, when they are planted for a trellis. When planted with care, they will begin to bear at the end of three years. The basket layers (Fig. 77) are prepared in the following manner. In the spring is made an osier basket (D) of an

15

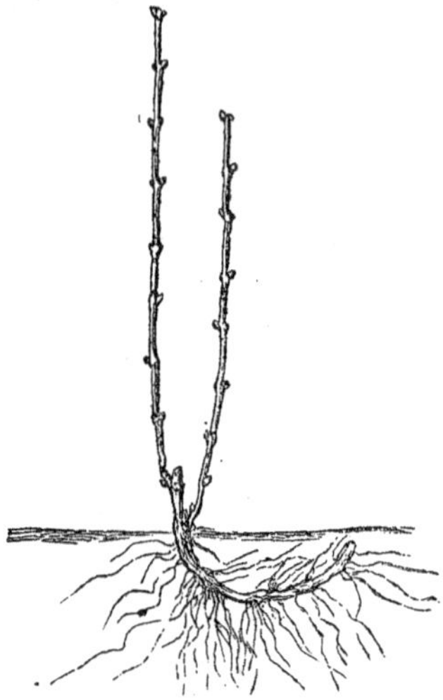

Fig. 76.

oval form and 12 inches long by 10 inches broad, and having a depth of 10 inches. These baskets should be made of green osier, that they may remain intact during a year. When the proper time for making the layer has arrived, the shoots to be operated upon being before chosen, a hole is pierced in the bottom of the basket at the point A, by which the shoot enters; each basket is then placed at a depth of 6 inches in the soil, and they are then filled with earth of good quality, to which has been added a portion of vegetable mold. Lastly, the top of the shoot is then cut in such a manner that only two buds or germs are left above the soil, and the whole is sustained by a prop. The operation is terminated by taking off all the buds on that part of the stem situated between the mother branch and the basket. This suppression is necessary to prevent these buds from absorbing the sap in their development at the expense of the layer. During the summer the two buds on the layer

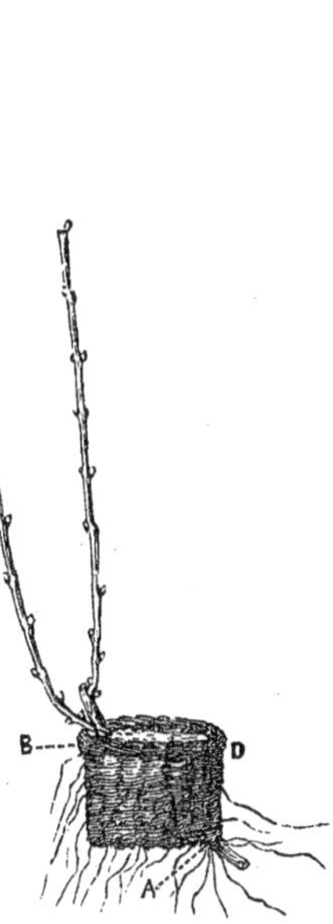

Fig. 77.

freely develop themselves, and put forth abundant roots, so that at the end of the year the layer is ready for use. The whole is then taken up and the layer hardly suffers at all from the separation from the parent stalk. This undoubtedly is the best method of propagation, and is that which is preferred at Thomery. Unhappily, on account of the expense attending the transportation of the basket layers, the cultivator is often compelled to use the unprotected layers, or *chevelées.*

Graft.—As to the graft, this mode of propagation, or multiplication, is employed for trellised vines only as an exception, and in circumstances analogous to those which render this operation necessary in ordinary vineyards. We have in the preceding part of this volume described the graft "*en fente bouture*" as one of the best for the vine. If, however, a *chevelée* may be used, it is to be preferred to any other for the purpose. It is planted near the vine, and the operation is conducted in the same manner as for the graft.

The great advantage of the *chevelée* is that it bears fruit the following summer.

An essential precaution, and one which is equally applicable to the three methods of propagation above mentioned, is the proper choice of the shoot intended to furnish the graft, the cutting or the layer. The shoot should have borne fruit during the year, and should be strong and in a healthy condition. The clusters should have been such as to exhibit in the highest degree the distinguishing qualities of the variety which it is desired to cultivate. Before the grapes are gathered, the shoots which appear the best suited for this purpose should be marked.

PLANTATION AND PROCESS OF BEDDING OR LAYING THE TRELLISED VINE.—*First Year.*—The superabundant moisture with which the soil is always impregnated during the winter is especially injurious to the roots of the newly-planted vine; it causes them to decay. The end of the winter, when the earth is drained sufficiently, is the time which should almost always be chosen for planting. There is no exception to this rule, but for dry and scorched soil like that of central and southern France. In such ground it is better to plant at the beginning of winter. The following is the process employed for layers in baskets:

If the land to be used is new, or if it has not lately been thoroughly cultivated, it should be dug during the winter to a depth of 32 or even so deep as 39 inches, if the soil is pebbly. The soil thus spaded up should extend to within 53.2 inches of the base of the wall. In the preceding chapters we have already spoken of the necessity that the soil should be such as to conduce to the health of the plants which it is intended to support. Such a soil is particularly essential to the vine. It may even be advisable, after the first spading mentioned, to carry it to a depth of 48 inches, and to widen it to an extent of 89 inches. The permeability of the soil should also be increased by the mixtures of earth already described, and the earth in all cases should be richly manured.

When the land is thus prepared, in the spring, a trench is opened 17 inches deep in dry, 20 inches deep in wet soils. The outer edge of this trench is 28 inches from the wall. The earth taken from it is deposited on each side. Vegetable mold, or compost, mixed with earth, is then spread over the bottom. In this trench the baskets containing the layers are placed. Should the soil be very dry the trench may be opened at 39 inches from the foot of the wall, instead of 28. A greater length of the stem is then bedded before it reaches the wall, and the roots, spread over a greater space, will more easily find the portion of moisture which they require. The space to be left between these layers is of course determined by that which it is intended shall be left between the vertical cordons upon the wall. If the cordons are intended to be 14 inches distant one from the other, the layers are separated by an interval of 28 inches, as after they have been laid, each layer should furnish two branches at the foot of the wall. A number of layers might be planted equal to the number of stems supporting the cordons, which are intended for the wall; but in that case they would be separated by a less interval, and would, as it were, starve each other. Moreover, the number of layers being greater, the expense would be increased.

It will, then, be more advisable to proceed in the manner just described; at all events, in those cases where the wall being only 39 inches in height, all the stems are to extend to the top.

When the stalks from the layers are separated at the base of the wall by an interval of 28 inches, the number of basket layers planted is equal to that of these stalks. If the first process is adopted, the layers are planted at the point A (Fig. 78), in the centre of the space by which the stalks against the wall are divided one from the other. In the second case, the layers are placed at the point A, in front of each of the points indicated by B.

The layers are planted in the following manner: From each

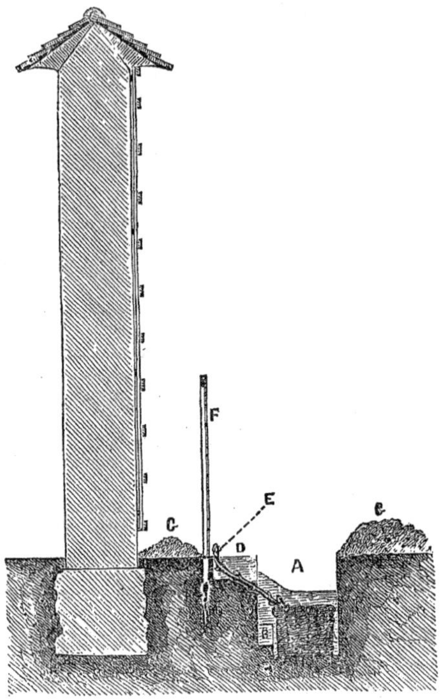

Fig. 78.

layer composed of two shoots, the least vigorous one is separated. The roots which issue from the basket are left untouched, provided that they are not broken, or dried by exposure to the air. This being done, at the bottom of the trench, and on that side which is farthest from the wall, a hole is made 6 inches deep and a little larger than the basket which it is intended to receive. In each of these holes a basket is placed in such a manner that the end of the shoot which it contains is turned toward the wall. That and the basket should be 10 inches below the level of the soil. A little notch is then made in the upper edge of the basket on that side nearest the wall, so that the shoot may be easily turned in the required direction. Then on that side of the trench nearest the wall, and in front of each basket, is made a smaller trench, as is shown in the figure at D, 3 inches deep and 10 inches long. In this the shoot is carefully laid, and it is filled with earth mixed with vegetable mold up to the level of the soil. The first trench is partly filled with the earth which was taken from it, mixed with vegetable mold. This operation is performed in such a manner that the trench is left empty to the depth of 8 inches, that the layer is buried to the depth of 10 inches, and that the top of the basket is covered by a bed of earth 2 inches in thickness. The operation is concluded by cutting off the shoot, as it leaves the earth just above the bud E, or that which is nearest the ground. The sap being thus concentrated upon a single bud, it attains a more vigorous development, and that part of the shoot which is buried, puts forth more roots, which pierce the bark with greater ease, in proportion as the leaf-buds from which they spring are nearer to the light. The end of the shoot above the earth is fixed upon a stave or prop 39 inches in length, and the remainder of the earth taken out of the trench is piled up on each side in the form of a shelving bank. The result of this last arrangement is to retain a greater degree of moisture in the neighborhood of the newly-planted shoot during the summer.

When the cultivator has no layers in baskets at his disposal, and is obliged to content himself with uncovered layers or even with cuttings, they should be planted in the same manner as the basket layers, only care must be taken to place the earth firmly around the *chevelées* and especially around the cuttings, and all that part which is under ground should be surrounded with earth which has been considerably enriched.

We will now proceed to describe the attentions demanded by this plantation during the next summer. When the bud E is developed, it is fixed upon the prop. As soon as it has attained a length of 20 inches the top is cut off; next the premature twigs which are thus developed are removed when they are 4 inches in length. The result this operation is to increase the size of the stem by limiting the evolution of the anticipatory shoots, and to accumulate in a small space all the nutritive juices taken up by the roots. It also promotes the increase of the roots along the newly interred layer. No bunch of grapes is allowed to remain on this shoot for fear of weakening it. The whole plantation should also receive two or three dressings in the course of the year. They should be applied, if possible, after rather a smart shower of rain, and when the earth has slightly drained. If the soil is light and dryness is to be apprehended, the trench and the little ditch should be covered with a bed of manure 6 inches in thickness, besides that which has already been applied, and finally, the trench is filled with the earth banked up on each side. After this operation, the whole appears like Fig. 78.

SECOND YEAR OF THE PLANTATION.—Toward the end of February, the shoot developed during the preceding year is cut at A (Fig. 79), above the three buds nearest to the base, then it

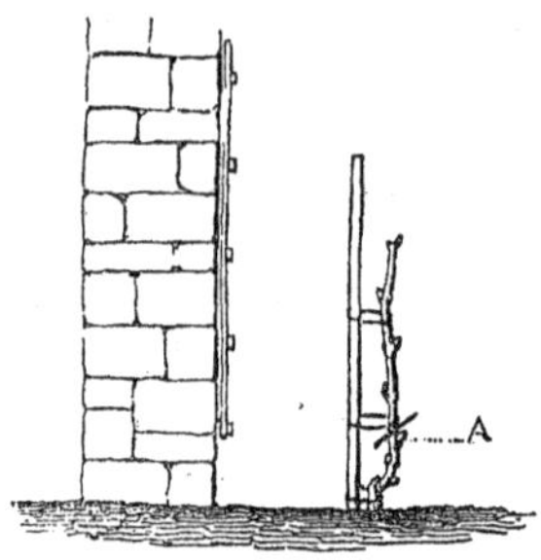

Fig. 79.

is attached to a prop 53 inches long which replaces the first. When the shoots have attained a length of 6 inches the laterals are pinched out, so as to preserve only the shoots from the three buds just described. These shoots are fixed upon a prop in proportion as they grow longer. They are not allowed to exceed the prop by which they are supported, and the process of nipping off the buds is continued. Should the shoots on the props be very vigorous, two clusters, at the most, should be left upon each, and should be treated in the manner which we will explain in the proper order. The same attentions are bestowed as in the preceding summer, and then a light dressing in November. The result then obtained is shown in Fig. 80.

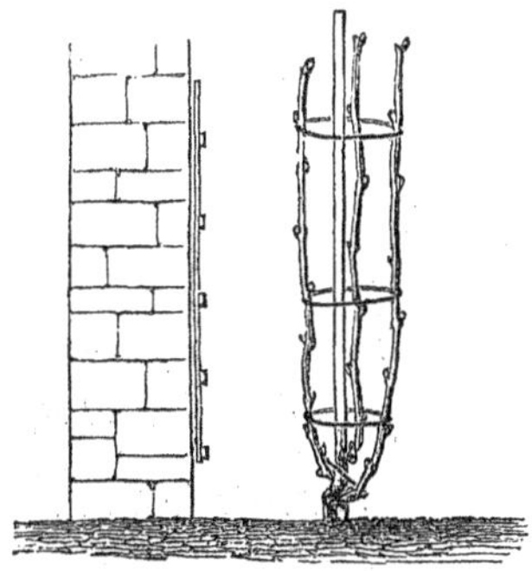

Fig. 80.

THIRD YEAR.—*Relaying.*—In good weather in the first of March, or, if in the South, in the autumn, the layers must be examined in order to know if they have put forth shoots sufficiently large and vigorous to be relaid. If uncovered layers, and still more, if cuttings have been planted, the cultivator will be obliged to wait till the following year and even to the year after to repeat the process of bedding or laying. The roots on the previously bedded shoot will not be sufficiently numerous, they would injure in their development the new layer which it is intended to put down, and the future health of the stalk destined to be placed against the wall would suffer. In that case only the two finest shoots of the young stalk should be preserved. These are cut to a length of only 6 inches, and upon these only a single shoot is preserved during the summer. Should they not be strong enough for relaying in the following year, the same operation is repeated. The stalks obtained from layers in baskets may almost always be rebedded from the third year. In that case the following method is employed. A trench 24 to 30 inches deep, according as the soil is more or less exposed to dampness, is opened at the foot of the wall, and is made wide enough to reach the young vines (Fig. 81). The

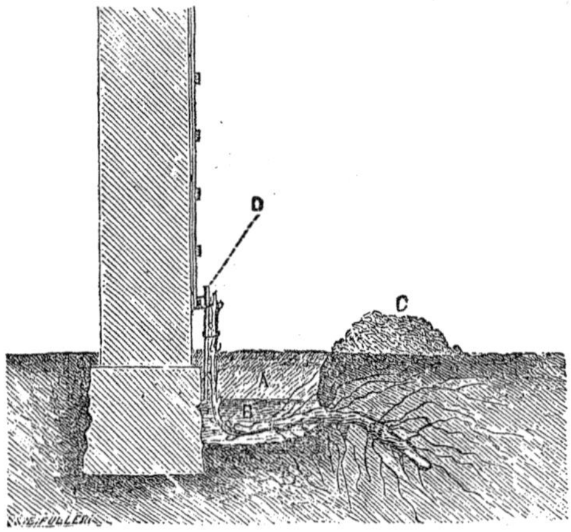

Fig. 81.

earth round the young vines is loosened with care until they turn naturally of themselves into the trench, in the bottom of which they are then placed in the manner shown by Figs. 81 and 82, that is to say, if each principal vine stalk is intended to produce two stalks to be trained on the wall (Fig. 82), the two

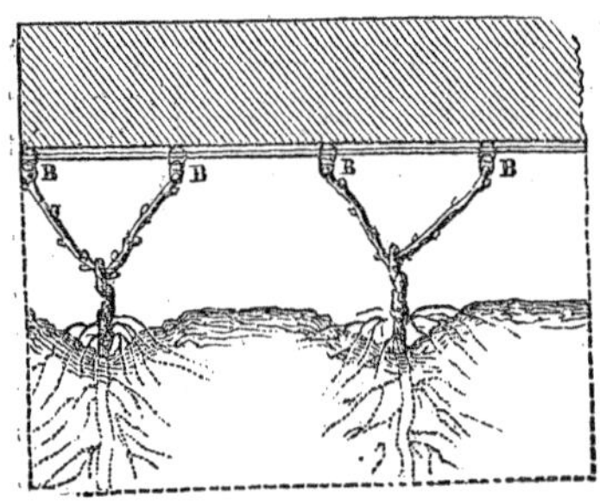

Fig. 82.

most vigorous shoots should be preserved, and they should be carried obliquely toward the wall, and from two stalks at the points B. If, on the contrary, it is intended that each principal stalk shall furnish but *one* stalk for the wall (Fig. 83), only

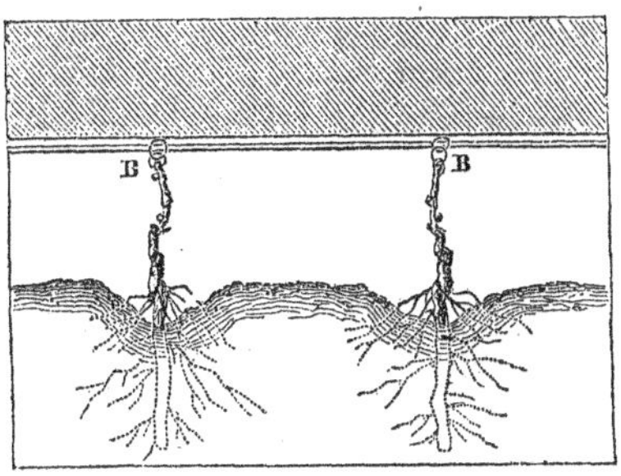

Fig. 83.

the finest shoot is preserved, which is buried in the trench and directed toward the wall at the point B, where it is intended to be trained. In both cases the shoots are covered as far as the foot of the wall by a bed of mixed soil and vegetable mold about 4 inches in thickness (Fig. 83). The trench is then filled with part of the earth which was taken from it, and the remainder is heaped up in a shelving bank at a distance of 40 inches from the wall, in order to preserve the moisture in the neighborhood of the newly laid vines, and thus facilitate a plentiful development of roots.

The upper extremities of the buried shoots are fixed at the base of the uprights of the trellis. These shoots are cut so as to preserve only the three buds nearest the base. This operation being concluded, the trellis presents the form shown in Fig. 81.

If the plantation of layers or cuttings has been in a trench at a distance of 40 inches from the wall instead of 28 inches, they must be brought to the foot of the wall only after a third laying, otherwise we should be obliged each time to cover too large a part of the shoot, which, as we shall see further on, will prevent them from properly taking root, and so injure the strength of the vine.

If this method of planting for the trellised vine is compared with that used in the majority of gardens, it will be seen that it is very different. In fact, the vines are almost always planted directly at the foot of the wall, and the only part buried is that which was originally below the soil; so that the vine, the roots of which ramify with great difficulty, cannot, when thus, planted, develop new radical organs upon the stems below the soil. It puts forth roots with great difficulty, it is long in recovering from its transplantation, and its vegetation is never vigorous.

On the contrary, by adopting the mode of cultivation used at Thomery, which we have just described, the vine is placed under much better circumstances. The first year there is buried, besides the stem first covered with roots, 10 inches of the shoot, which during the two or three years preceding the relaying, covers itself with vigorous roots. Two or three years after this, 14 inches of the shoot are again laid, which in a little while is completely covered with roots. Each stalk intended for the wall is then provided with an underground stem 44 inches in length, bearing through all its length numerous and vigorous roots, which give to the vine more strength and hardiness than is possible when the method of which we spoke first is used. When uncovered *chevelées*, or those in baskets are used, the cultivator may be tempted to lay at once a length of shoot sufficient to bring the upper end directly to the foot of the wall, a length, for instance of 24 inches. This is a very bad plan, for the stems do not properly take root only upon the 12 to 14 inches nearest to the upper ends, because the woody and cortical fibres which run down from the buds to produce roots are not sufficiently numerous to put forth roots enough, and they pierce the bark at the same time that they meet the soil. It is desirable to lay only 14 inches at the most, if it is intended that the underground stalk shall be fully provided with roots throughout its whole extent.

METHOD OF PRUNING ADOPTED FOR THE VERTICALLY TREL-

LISED VINE WITH OPPOSITE SHOOTS.—CONSTRUCTION OF THE FRAME.—*First Year.*—The shoots having been laid and brought to the foot of the wall, the buds are watched in their first development to see that they are not harmed by caterpillars, snails, or other destructive insects. When the three shoots have attained a length of about 6 inches, the *stipulary shoots*

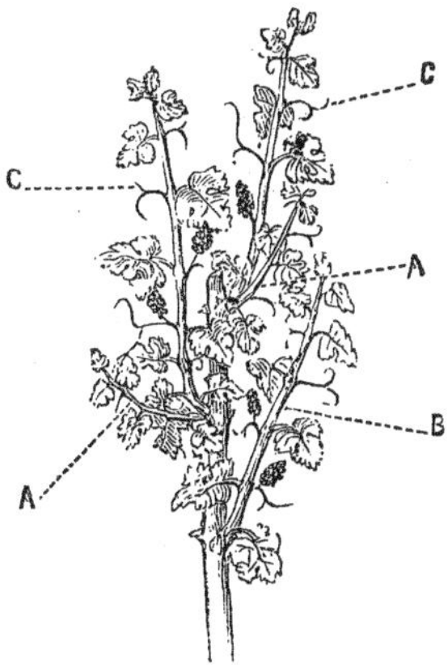

Fig. 84.

(A, Fig. 84), which often grow by the side of the shoots properly so called, are taken away. Then, when they are about 12 inches long, we begin to break the tendrils which uselessly absorb the sap. This breaking of the tendrils is continued through the period during which the length of the shoot increases, and should be put in force while the tendrils are yet so soft that they may be easily broken. That is also the time which should be chosen to begin the formation of the stalk intended to be trained upon the wall. The following is the method then employed:

Let us suppose that one of these young stalks is represented by Fig. 84. From the three stalks which have been preserved, one is chosen having a leaf 12 inches above the ground. Let us suppose in our figure that this is the second shoot on leaving the ground, and that the leaf aforesaid is situated opposite the second cluster. This cluster is taken off and the shoot is cut immediately above this leaf, as in B, Fig. 85. The top of the two

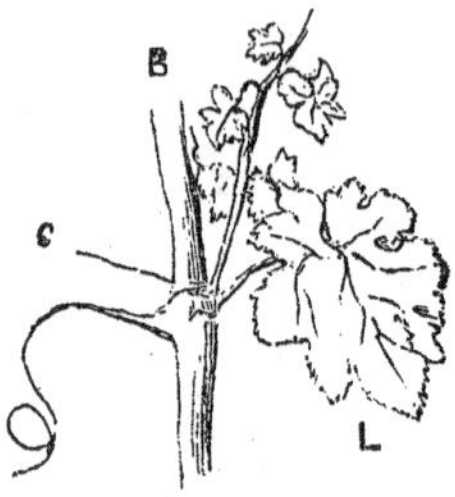

Fig. 85.

other shoots is then removed in order to hinder them from too great a growth to the detriment of the shoot upon which it is intended to operate. We may then proceed to train it upon the frame. The shoot under treatment is placed in a vertical position, and the two others are attached at an angle of forty-five degrees. A premature stipulary shoot will be seen immediately to spring from the axil of the leaf of the cut shoot (A, Fig. 85). This shoot should be broken when only an inch or two in length, so that the bud B at the base of this shoot is forced to develop itself. Before long this bud gives birth to a shoot

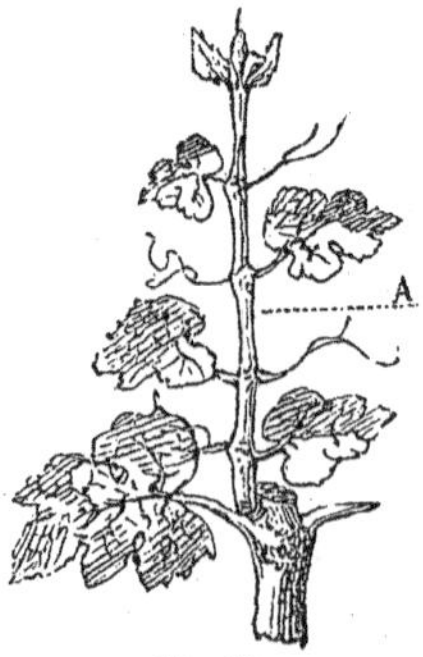

Fig. 86.

(A, Fig. 86), which is allowed to grow, and which is trained vertically. These young main stalks require no other care during the summer, so far as the frame is concerned, than the complete suppression of all the premature stipulary shoots

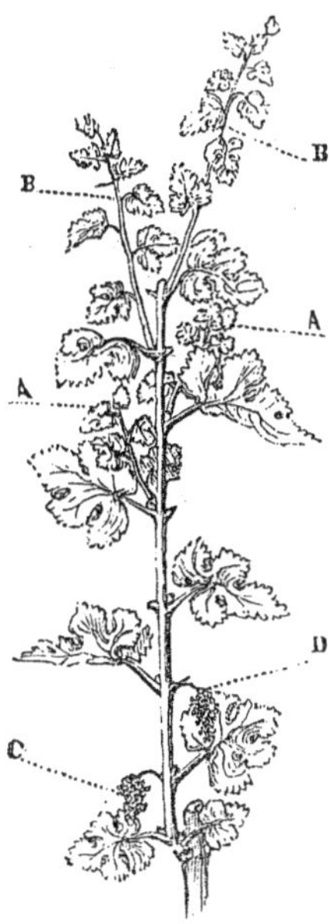

Fig. 87.

(A, Fig. 87), or of premature shoots commonly so called, as also of the tendrils. Upon each shoot should be left only the clusters C and D and the primitive leaves. These attentions should be given each year to all the shoots preserved.

Second Year.—The stems operated upon in the manner just described present the appearance of Fig. 88. They are then subjected to the second pruning. The two shoots (B) are completely taken off by cutting the first at A. Then the premature shoot C is cut at D immediately above the bud situated near the

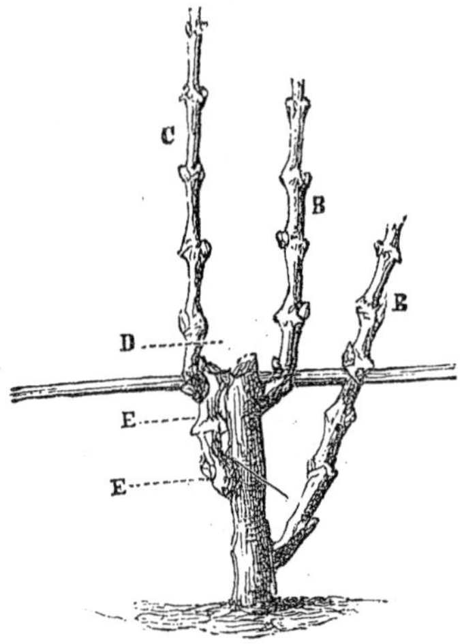

Fig. 88.

base. During the following summer this bud develops itself as well as the germs immediately below it upon the secondary shoot,

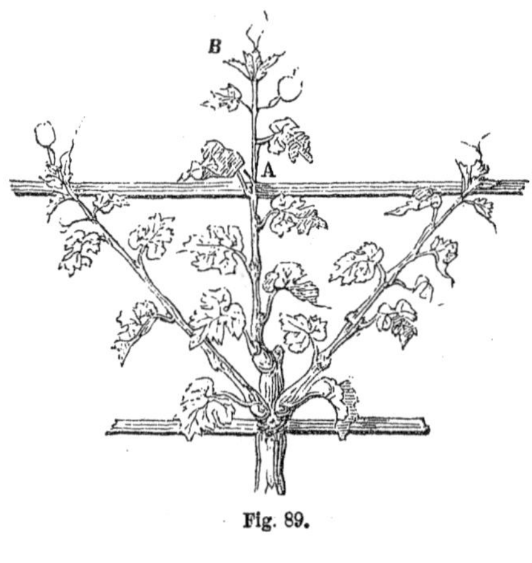

Fig. 89.

indicated by the letters E E, which is called the *spur* (talon). The number of buds on the spur may be three or four. But two buds on the spur, one on each side and one at the top of the shoot, are preserved. The produce of the buds E E is entirely removed. This last operation is performed as soon as the shoots from the spur have attained a length of 4 inches. When the remaining shoots are fastened to the frame, the young vine presents the appearance of Fig. 86. When the centre shoot (B) puts forth, as it increases in length, a leaf above that point where the first pair of lateral shoots is attached, it is cut *above* this leaf at the point A, as in the preceding summer, in order to obtain from the axil of this leaf a new shoot for the formation of a second pair, which must be treated in the same manner. The two lateral shoots are subjected to the same operation.

Third Year.—In the following spring each stem on the walls

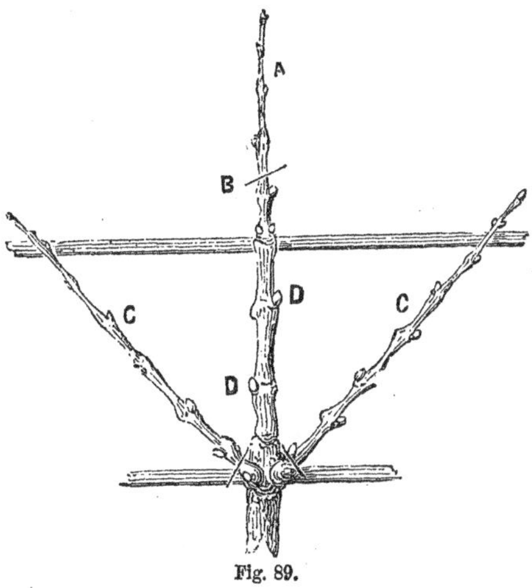

Fig. 89.

presents the appearance of Fig. 89. The shoot A is cut at the point B, in order to obtain the same result as in the preceding year. As to the branches C, they are cut near their base in order to form the two first *coursons* or double branches shown

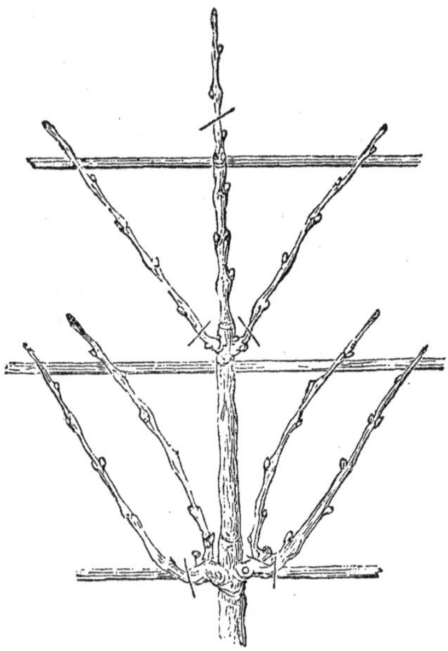

Fig. 90.

in Fig. 90. The same development takes place during the summer below the point B, as well as the same operation upon the new terminal shoot. The product of the buds D is removed.

Fourth Year.—Fig. 90 shows the result of the operations performed during the preceding years. The same method of pruning is practised one year after another until the trellised vine has covered the space for which it was intended, when it presents the appearance shown in Fig. 74.

All that we have just said applies to those stems which rise to half the height of the wall. Those which extend to its top grow more rapidly during the first years. During the summer, after the layering by which they have been brought to the wall, two shoots are left upon each of the three first shoots. The following year, at the winter pruning, the strongest of the three shoots resulting from thence is chosen; the two others are taken away and the remaining one is cut at 20 inches above the point where it is attached to the frame. In summer it is allowed to retain but three buds, which give place to three new shoots. The best of these is again chosen and extended also to 20 inches. The same process is repeated till the vertical stalk reaches the point where it is intended to support lateral branches. Then the same series of operations is employed as in the first case.

This method of forming the main stalks has this advantage. that each pair of lateral branches being separated by a regular interval of 10 inches and by a knotty place at the point of attachment of the successive extensions, the course of the sap is arrested below every one of these knots and thus obliged to act with the same intensity on all the lateral branches of the same stalk. Such is not the use in the vertical cordons which are more rapidly formed, as they are more extended at each pruning.

Care necessary to the Lateral Branches.—*First Year.*— The essential principles of pruning the lateral branches are the following: In the case of the vine, the clusters are attached to

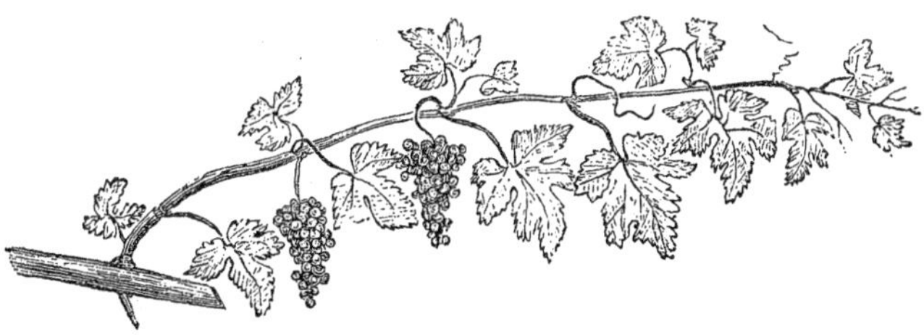

Fig. 91.

shoots proceeding from the branches of the preceding summer (Fig. 91). The shoots accidentally developed on the old wood never bear grapes (Fig. 92).

The further the buds are removed from the base of the branch, the more fruitful are the shoots to which they give rise.

Fig. 92.

Hence it appears that the shoots should be left entire, or be left very long. But in that case we immediately encounter the following inconveniences. Thus, if the shoot in the Fig. 93 is cut in B, the buds C and B are the only ones which will be

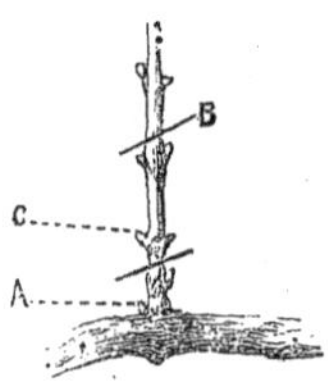

Fig. 93.

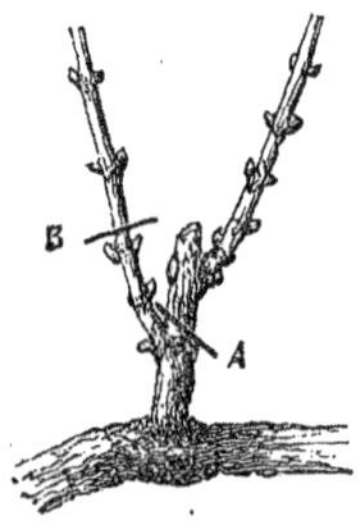

Fig. 94.

developed, and we shall have in the following year the result shown in Fig. 94. If, then, we trim the shoot at the points A and B (Fig. 94), we shall have two new shoots produced at the top of the shoot B. Continuing to trim in this manner the lateral branch or immediate support of the young shoot increases in length each year from 4 to 6 inches, and thence results great confusion through the whole extent of the trained vine, and moreover, a progressive enfeeblement, or, as it were, starvation, of the new shoots, and, consequently, an immediate diminution of fruitfulness.

On the other hand, if the shoot in Fig. 93 is cut so as to

preserve only the bud A, this bud is so near the old wood that the shoot produced from it will bear no grapes.

It will be best, then, to cut this shoot (Fig. 93) as short as possible, to hinder the lateral shoot from increasing in length, but in such a manner, however, as to preserve a bud far enough from the old wood to produce grapes. Experience has shown that in order to attain this double end, the shoots from varieties of only a slight or average degree of strength should be cut above the two buds the nearest to the base, one of these two being that bud which, hardly visible, is on the base of the shoot itself—that is, just where it springs from the stalk (Fig. 93). Two new buds are developed, and in consequence, two new shoots. The branch will then present the appearance shown in Fig. 95.

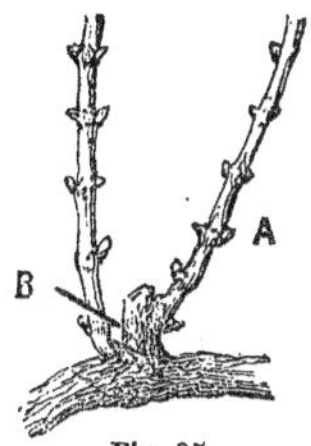

Fig. 95.

The shoot A has borne clusters during the summer. The shoot B is too near the old wood to have produced anything. It is called *the shoot of replacement*—that is to say, it is that intended to undergo the next pruning. For that, almost all the old wood is cut from the top of the spur. Then the shoot B is cut above the two buds nearest its base. During the summer two new shoots are thus produced, and each year the same method of pruning is repeated, so as to allow the old wood to increase as little as possible in length, and keep the fruitful shoots as near as possible to the direct channel of the sap. Such is the method of pruning applied to the branches intended to bear grapes for the table.

There are, nevertheless, varieties so hardy that, should they be subjected to this process, no fruit, or very little, would be obtained. The different varieties of muscats, the Frankenthal, and others which we have noted in our list, are of this description. For these, the shoots should be left a little longer. They are cut off below the third bud. This difference does not result in increasing the length of the lateral branches. In fact, such is the strength of these vines that three shoots are obtained from each lateral branch. That from the top, which generally bears the clusters, is the one preserved, then that at the base, intended to undergo the next year's pruning. The intermediate one is suppressed. The same operation is each year repeated.

Disbudding the Lateral Branches or Coursons.—When the *coursons* are cut so as to preserve but two or three buds, it will often happen, nevertheless, that a larger number will be developed. Only two, at the most, should be left at each point. The shoot A (Fig. 96), nearest the old wood, is preserved as a shoot of replacement, together with that farthest from the same

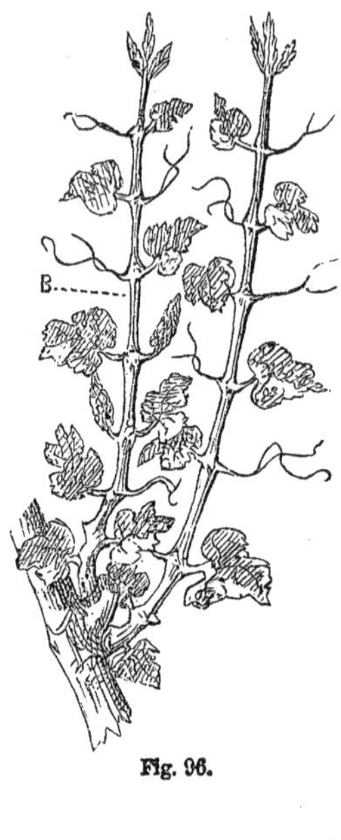

Fig. 96.

point B. The latter generally bears the clusters. There are, however, two cases in which but a single shoot should be left on the *courson*. First, when none of the shoots of the *courson* bear clusters; then a single shoot, that from the base, is useful as a shoot of replacement. By the others being suppressed, the remaining one becomes stronger and will yield finer fruit in the following year.

Second. When the two shoots of the *courson* both bear clusters, which occasionally occurs in very fertile years. As it is advisable to leave only two small clusters or one large one to be supported by each *courson*, as we will presently explain, a retrenchment will be necessary. In this case, the shoot from the base only is preserved, and it will become at the same time a fruit-bearing shoot and a shoot of replacement. In consequence of this retrenchment the shoot in question will acquire more strength, it will bear better grapes, and the new shoot will afford the finest products of the following year.

The proper time for putting in practice these different trimmings, is, as soon as the young clusters make their appearance upon the shoots, that is to say, when they are about 10 inches long. We must repeat what we have said concerning the cutting of the shoots—that there should be left upon each one of the shoots preserved only the clusters and the primitive leaves. Then all the supplementary shoots and the tendrils should be removed as soon as they appear.

PINCHING THE SHOOTS.—The buds on the shoots of the vine, as on those of other trees, should often be pinched back. The end of this operation is to prevent the shoots from confused growth, to diminish the growth of some of the shoots to the profit of feebler ones, and finally to favor the development of the grapes by enabling them to profit by the sap, which would otherwise pass to the shoots which would spring from the buds destroyed.

In order to obtain these different results the buds on the shoots should be pinched off as they develop themselves to the length of from 16 to 20 inches, and their extremities only should then be cut.

Manner of fastening the Shoots in Summer.—The shoots of the vine are fastened in order to prevent their being broken by the wind, and in general this fastening should be twice practised upon the same shoot. The first fastening is made when the shoots have attained a length of about 12 inches. Then the shoots are but slightly compressed in the rush which serves as a ligature. Otherwise, in growing, they would break themselves.

Fifteen days after this first fastening, we proceed to the second, or *récollage*, as it is called by the cultivators of Thomery. At this time the shoots are tied as close as is necessary to arrange them conveniently. This process of fastening should be successively made for the different shoots of the same vertical main stalk, and by beginning with the most vigorous we may equalize their strength. The shoots of the vertical cordons should be inclined at an angle of forty-five degrees.

Renewal of the Coursons.—We have seen that in spite of the care which has been taken to keep down the spurs by an annual trimming, to the shoot nearest the base, they will always increase a little in length, and the shoot which they bear will diminish in vigor in proportion as they are removed from the point where the spur or lateral branch is attached to the cordon or vertical main stalk. In order to remedy this inconvenience, the shoots which sometimes grow at the base of the spurs are carefully preserved, whatever may be the age of

Fig. 97.

the spurs from which they spring. Then, of the two upper shoots, that which bore the worst cluster is suppressed. The following year the spur is cut at A, Fig. 97, and the shoot B is cut above the two lowest germs or eyes in order to form a new *courson* or spur.

REPLACEMENT OF THE SPURS (COURSONS).—Sometimes also certain spurs disappear entirely or are not developed where they are expected, and in either case spaces are left which it becomes necessary to fill. This accident may be remedied by the graft.

CARE OF THE GRAPES.—It is in particular the intelligent labor bestowed upon the grapes from their first appearance to their maturity to which the cultivators of Thomery are indebted for their success. The following are the processes adopted:

SUPPRESSION OF THE SUPERFLUOUS BUNCHES.—Too large a quantity of grapes upon the vine produces the same result as a superabundance of fruit upon other trees. A great quantity of grapes are gathered, but the clusters and the berries are small, and the vines are enfeebled for the following year. If the necessary retrenchments are made, the same result in weight is obtained, and the grapes are larger, better flavored, and command a higher price.

THINNING THE BUNCHES.—When the berries have attained the first stages of development, it will be proper to thin them.

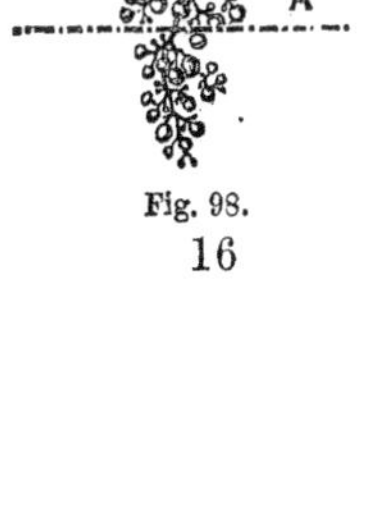

Fig. 98.

With a straight, pointed pair of scissors we cut from each bunch—first, all the abortive berries; and secondly, those in the middle of the bunch, together with some of those which, although on the outside, are too much crowded. If the bunches are very long, as is often the case with young and vigorous vines, the point of the bunch (A, Fig. 98) must also be removed, since the berries which it bears would be slow in ripening. The result of this thinning is, that, other things being equal, the grapes are ripe fifteen days earlier, the berries are a third larger, and those intended to be kept through the winter will keep better.

The thinning practised at Thomery is performed by women, and is applied to at least half the harvest—that is to say, 500,000 kilogrammes of chasselas.

GATHERING THE LEAVES.—At the time when the thinning takes place should also be applied the first *épamprement*, or picking off the leaves. At first only the leaves turned toward the wall and those more or less broken or distorted are removed. When the berries begin to look transparent, a second *épamprement* takes place. A few leaves on the front of the vine are then removed in situations where the foliage is thick; but the leaves which shelter the branches, *the parasols*, are preserved with care. Finally, when the berries are entirely cleared, and begin to turn yellow, the leaves which shadow them are removed. If they are exposed earlier the berries will harden and cease to increase in size. The bunches thus uncovered are exposed alternately to the dew and the sun, by the action of which they acquire that beautiful pale yellowish brown which distinguishes the chasselas of Thomery.

Black grapes require particular care in this respect. The first removal of the leaves should not take place till the grapes are completely colored.

These successive strippings of the leaves from the vine result in progressively arresting the annual growth of the vine, a long time before it would otherwise cease. The fruit, therefore, sooner begins to mature, and will be completely ripe by the first cold weather.

Protections.—The very projecting copings which we have recommended for trellised vines are insufficient, if the wall is more than 80 inches high, to protect the grapes from the dampness of the atmosphere. It will then be advisable to place a movable pent-house at about half the height of the wall after the last gathering of the leaves in the middle of September. This pent-house should project about 20 inches.

Annular Incision.—Refer to page 234 for the description of this operation, intended to hasten fifteen days the ripening of the grape, and which will increase also fully a third the size of the berry.

Renewal of the Trellised Vine.—The trellised vine, attended in the manner we have described, will bear fruit for more than fifty years. But there comes a time when the successive renewal of the spurs produces upon them so many knots that the circulation of the sap is interrupted. The vegetation becomes languishing, many of the *coursons* wither, and the vertical stems themselves finally perish. When this state of decrepitude first manifests itself, the cultivator proceeds to the renewal of the vine. All the vertical stalks are cut at about 8 inches above the soil (Fig. 99). This trimming concentres the action of the sap upon this point, and so develops a certain number of shoots. During the summer the most vigorous are chosen and the others removed. The following year the reserved shoot is cut above the third bud, and the same care before described is applied to the three resulting shoots. Then the process is continued as for the establishment of a young vine. To assure its success, it is well to remove, from the time when the shoots are suppressed, as much earth as possible from the foot of the trellis without injuring the roots of the vine, and we should apply abundance of manure, which should be covered with a bed of new earth nearly equal in thickness to that removed. When the trellis to be renewed is in a state of advanced decrepitude, and when a certain number of vertical stalks are completely withered, and the regularity of the whole is lost, we proceed in a different manner. Each vertical stem is cut off, as we have said, above, and those which are dead

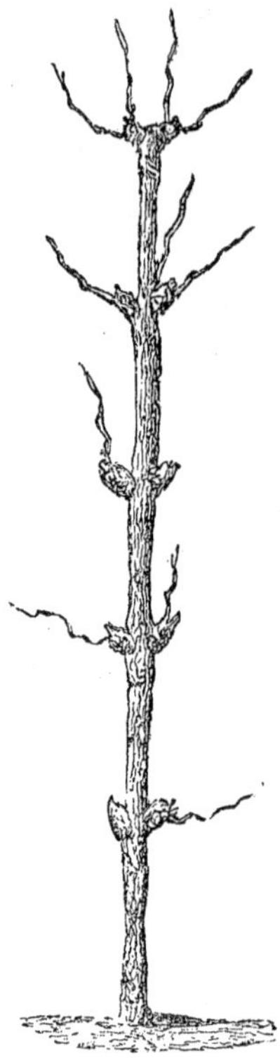

Fig. 99.

removed. During the summer the two most vigorous shoots on each vertical stalk are preserved, and they are allowed to grow to the top of the wall. The following year there is taken away from the foot of the trellis as much earth as possible, about 16 inches, taking care not to harm the old roots. The earth is hollowed out, completely as it were isolating the base of each vertical stalk. Then they are laid at the base of the trellis previously arranged for their reception. As each one leaves two shoots, and as this number is more than sufficient to furnish the required number of vertical stalks, we preserve only the proper number, choosing the most vigorous for our purpose. These stalks and shoots are finally extended on the ground by means of wooden hooks, in such a manner that the new shoot directed toward the wall leaves the ground at exactly that point where the new vertical stalk should rise. A bed of manure, 3 inches in thickness, is then spread, and the rest of the hollow is filled with new earth. All these vertical stalks will develop with exceeding vigor during the year, and will then be managed like those of a new plantation. We saw thus renewed, in 1848, a trellis more than eighty years old, belonging to M. Rose Charmeux. The operation was attended with no difficulty, and its success was complete.

It will readily be perceived that by the aid of this renewing process the duration of the trellised vines is almost indefinite, and it will seldom be necessary to replant. The cultivators of Thomery have a proverb, "He who plants an *espalier* is not there to take it away." This mode of a renewal may be applied to an old trellis more or less regularly disposed in horizontal cordons, which it may be desirable to replace by vertical ones. The process in such a case is as follows:

In the spring each cordon is cut immediately above the spur (*courson*) nearest the base (Fig. 100). During the summer two

Fig. 100.

shoots are preserved upon each spur and allowed to grow freely. The following year the ground at the foot of the trellis is dug out as we have explained. Then the foot of each vertical stem is deeply laid bare and laid down horizontally so that the extremities of the shoots are connected to the foot of the wall at each of those parts were it is intended they shall form new vertical stems. The rest of the process is conducted in the manner already described.

CULTURE OF TABLE GRAPES IN THE OPEN AIR (PLEIN VENT).

The table grape is also cultivated in the open air,* but the climate of Paris is the extreme limit of this culture. The vines are arranged upon *espalier* and then managed as before described. They are even sometimes trained upon poles or stumps, and the method pursued is then the same as for the ordinary vineyard.

At Thomery the interval which separates each inclosure is used in the following manner: *Espaliers* are established parallel to the walls. The first is at 80 inches' distance, and the others are separated by an interval of 8 ft. 6 in. These *espaliers* are sustained by a trellis similar to that on the wall. They are supported on wooden posts, or, as is better, on those of schistose stone, analogous to slate. These posts are placed at a distance of 5 ft. 4 in. one from the other. Sometimes for these posts are substituted iron uprights fixed in prisms of sandstone placed in the ground. In this case the wooden cross-pieces may be replaced by lines of iron wire which pass across the uprights. The main stems of the vine form upon this frame a series of little vertical cordons like those just described. These *espaliers* are, moreover, planted with the same care as the trellised vines, and are treated in the same manner.

The interval of 8 ft. 6 in. which separates each *espalier* is occupied by a row of vines on poles, propped up as in the

* "Open air" is here used not in contradistinction to vines protected by glass, but those simply protected by walls and copings as just described.

ordinary vineyard, and subjected to the same method of cultivation. These poles, separated by an interval of 53 inches, rise to a height of 13 inches above the soil, so that the rain may not cover the grapes with mud.

In the same climate, the same variety of grape supported on a pole is always inferior to that cultivated upon a wall. The grapes from the pole vines are always worse than those from the *contre espalier.*

The earliest varieties only should be cultivated in this manner, since the temperature of the *contre espaliers* is always lower than that of the *espaliers.*

CULTURE OF TABLE GRAPES IN SOUTHERN FRANCE.

In the south of France the greater warmth and dryness of the climate hastens to a great extent the annual vegetation of the vine, and the ripening of the fruit is accomplished without its being necessary to increase the warmth of the atmosphere artificially, or to moderate and even to arrest the growth of the vines. Hence the vine grows most vigorously and the choice varieties of table grapes which are native to these regions have a much greater development than those which belong to the centre and north of France. Finally, these varieties require less pruning in order to produce grapes. These different considerations give rise to the following modifications in processes of grape culture for those regions.

First. The vine should be placed on *espaliers*, single or double, the supports of which should be like those already described. In all cases the walls of the garden which have the warmest exposure should be devoted to the vine, and for these walls should be selected the latest varieties.

Second. The vines should be planted before winter, as if planted later they suffer much from the dryness of the spring.

Third. As the vine grows with much more strength in the south than in the north of France, whether on account of the climate, or the nature of the varieties peculiar to that region, it is necessary that they should be planted at a greater distance one from the other. For the vertical cordons, with opposite lateral shoots, it will be proper to leave an interval of 24 inches between each cordon instead of 14.

Fourth. The *coursons* of those varieties analogous to the chasselas, on account of their strength are cut so as to leave two buds as we have explained, but all those which grow with more strength are cut so as to leave *three* buds.

Fifth. The operation of thinning the clusters is as efficacious in the south as in the north, but removing the leaves would be much more injurious than beneficial. Only the leaves which cover the clusters are to be taken off, and those only at the time when the grapes are perfectly transparent.

Sixth. The vine in the south being stronger than in the north, a third more clusters than the amount previously specified are allowed to remain upon the vine.

Diseases of the Vine—Destructive Animals and Insects.—The diseases of the vine have been already referred to, and we will confine ourselves at present to the consideration of the destructive animals and insects, which especially attack the trellised vine.

Birds, and particularly sparrows, thrushes, grossbeaks and black-birds are the great enemies of the trellised vine. When these birds do not fly in large flocks and descend in great numbers upon one place, they occasion little mischief, and the cultivators of Thomery adopt no precaution against them. Nets undoubtedly would be a good defence, but their price prevents their being employed over a large surface.

M. Orbelin, of St. Maur, near Paris, has contrived, as a defence against birds, little mirrors with a double face, of a very moderate price, and the result, up to the present time, has been very satisfactory. In the spring the first young shoots are often devoured by snails or slugs. Their size, their slow progress, and their habit of taking refuge in the chinks of the wall or behind the trellis, and of coming out in the morning or during the rain renders their destruction easy.

The *kermes*, known also under the name of gall insect, belongs to the genus *coccus*, and particularly attacks the peach and the vine. When it has acquired complete development toward the end of May, it presents the following appearance:

The male (A, Fig. 101) appears in the form of a little multipede

Fig. 101.

or woodlouse covered with white dust. The female appears like a little brown shell, B, adhering very firmly to the branches of the trees. About this time the males impregnate the females and die. The females lay their eggs directly, and the eggs remain surrounded with a little mass of white down, and covered with the dried body of the female, who expires as soon as they are deposited. These eggs hatch rapidly, and the insects issue from the shell which covers them, toward the end of June, to the number of more than a thousand. Hardly visible to the naked eye, they spread themselves over the surface of the leaves and young shoots, and destroy them by piercing their epidermis and absorbing their fluids.

Toward the month of November, when the leaves fall, the kermes abandon them and fix themselves on the branches, choosing in preference, where the trees are *en espalier*, the side next the wall, where they remain torpid through the winter, appearing like little brown stains. In the month of April they change their skins, rapidly increase in size, and give birth to a new generation.

The measure-worm is the larva of a moth, which in the spring greatly injures the vine by devouring the young shoots as they are put forth. It is difficult to find it, as it has the form and color of a little dried stick. It carries on its ravages during the night, and it is then that the cultivators of Thomery, armed with lanterns, seek it out and destroy it.

GATHERING AND PRESERVATION—FRESH GRAPES.—The grapes should be gathered only when perfectly ripe. The longer the

Fig. 102.

vintage is delayed in the centre and north of France, the higher is the flavor of the grape. The first frosts of autumn, to which it is very sensitive, should however be anticipated. The gathering should take place in a dry time. Each cluster should be taken by the stem, and detached by means of the pruning shears.

As the grapes are gathered they are deposited in little baskets lined with vine leaves and fern. These baskets are arranged on what is called a crotchet, or sort of hod, shown in Fig. 102, which can be carried by one man to the storehouse, or to the place where the grapes are packed for market.

The following is the manner employed each year in the preservation of a great quantity of grapes by the cultivators of Thomery:

First, a certain portion is retained on the trellis to the latest possible moment. They choose the clusters from the two upper cordons of the walls having a southern exposure. These grapes are the least watery, and consequently the least susceptible to cold. They guard them by sheltering them with leaves of fern, and even with straw matting, and thus preserve them until Christmas. The grapes which they wish to preserve still later they treat in the following manner: Those which they wish to retain till May are chosen from the poles, or the counter espaliers. The bunches are taken which have been subjected to the thinning process and which are formed of the largest and least crowded berries. They are cut a little before they are completely ripe--that is to say, from the 25th of September to the 15th of October. The grapes intended to be kept only till March, may be taken from the espaliers, and are gathered from the 1st to the 15th of November.

The place where the grapes are kept is generally some room or building connected with the house, and especially devoted to this use. Shelves about 30 inches wide, placed one over the other, cover the walls from floor to ceiling. In the middle of the room, and 30 inches distant from the lateral shelves, another series of shelves rises to the ceiling. These shelves are composed of a frame of wood filled up with a grating of iron wire. It is upon this grating, which is covered by a slight layer of very dry straw, that the grapes are spread. They should often be inspected, and the berries which begin to decay should be removed by the scissors.

A storehouse on this plan presents the following inconveniences. Heat must often be introduced in order to defend it from the winter's cold, and the result is an injurious change of temperature. On the other side, the accumulation of moisture makes it necessary that it should be aired from time to time, and produces the same result in an inverse mode. Finally, if the currents of air produced by this ventilation are too great, the grape dries, shrivels, and loses, if not its quality, at least its commercial value. We think, then, that it is better to use the storehouse a description of which the reader will find at page 685 of the second part of this work. It will be necessary but to change the arrangement of the shelves, and also to use chloride of calcium with precaution, for fear of shrivelling the grapes.*

When it is necessary to preserve only a small quantity of grapes, the same storehouse will serve at once for grapes and

* The reference here is to the "Cours Elémentaire d'Aboriculture," from which the present account of the Thomery system is translated. M. Du Breuil there gives a very full and accurate description of a room or house for preserving fruit of all kinds; the principal features of which are the provision of means whereby the fruit is kept at an equable temperature, free from all pressure produced by the fruits pressing upon each other, and free from dampness. The latter point is attained by keeping a vessel of *chloride of calcium* in the house—a substance which must not be confounded with *chloride of lime*, which would quickly destroy the fruit. This caution is not unnecessary, as it is only a few years since a writer in the "Horticulturist" recommended chloride of lime for the purpose; having, no doubt, used this term under the impression that it was *simpler* than the word calcium. Chloride of calcium may be purchased cheaply, or it may be made by dissolving chalk or lime in hydrochloric acid. It must be evaporated to dryness, and calcined at a red heat; after it has become moist by exposure to the air in the fruit room, it loses its power of absorbing moisture, and must be again dried and calcined, but after undergoing this process it is as good as new. Most cellars in American dwellings maintain a very equable temperature during winter, and it has occurred to us, that a small wooden press, made air tight, shelved and kept dry by means of chloride of calcium, would form no bad substitute for Du Breuil's "Fruiterie." We hope to try it next season.

other fruits. The grapes should then be spread on shelves by themselves, or can be arranged in the following manner, which has the advantage of economy of space. Each bunch should be suspended by the point on a little hook of iron wire in the

Fig. 103.

form of an S (Fig. 103). Thus attached, they will be less liable to decay, because the berries will have a tendency to fall apart

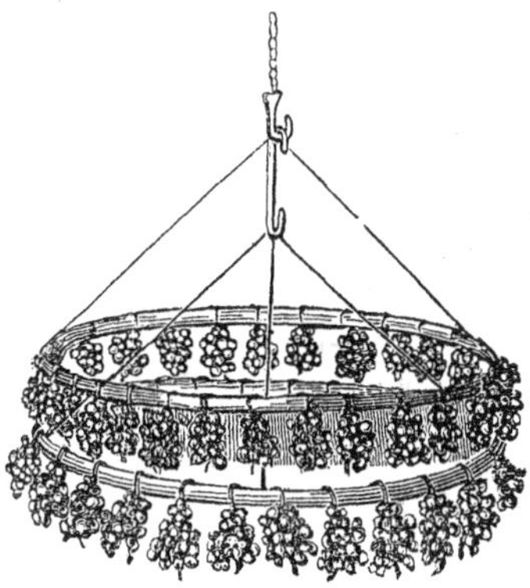

Fig. 104.

from each other. The bunches are then suspended by the upper hook of the S, around hoops hung one over the other (Fig. 104), and themselves suspended from the ceiling of the room, and moved up and down by little pulleys. If we should wish to preserve a larger quantity of grapes, we may, for the sake of economizing space, substitute for the hoops wooden frames in

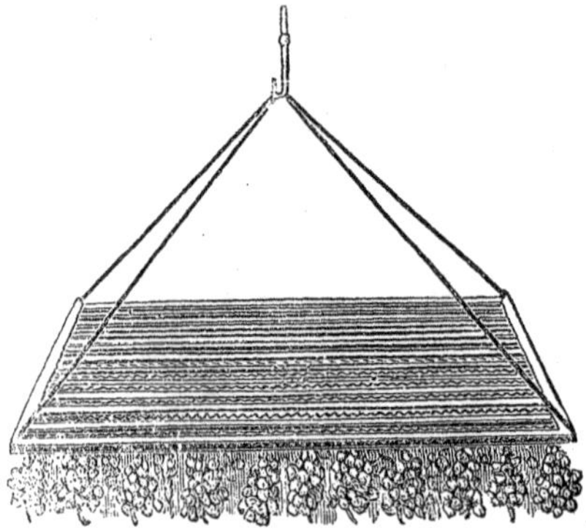

Fig. 105.

the form of sashes, as shown in Fig. 105. These sashes are furnished with rods, separated from each other by an interval of 4 inches, and having on one side little points intended to receive the hooks by which the clusters are suspended. These sashes are hung from the ceiling in such a manner as to fill the entire space, and like the hoops, move up and down. However, the grapes thus preserved wither and lose more of their quality than those preserved upon shelves.

Dried Grapes—Raisins.—The large proportion of saccharine principle which the grapes of the south generally contain, renders it easy to dry and preserve them. They have thus become the object of special attention and considerable commerce for some countries in the south of Europe where are cultivated the varieties best adapted to this purpose. We have noted the most desirable of these varieties in our list. Malaga, Calabria, Egypt, and Roquevaire in Provence are the principal places devoted to this culture. Zante in particular is distinguished for the Corinth grape, or currant.

The process most commonly employed for the preparation of raisins is the following:

When the fruit approaches maturity, the stem of the bunch is twisted, and the leaves are removed in part from the branch in order to expose the grapes to the influence of the sun's rays, in order to favor the action of the essential principles and diminish the superfluous moisture. The grapes are gathered at the proper time, and the spoiled berries are carefully removed.

After which the clusters are left upon hurdles exposed to the sun for one day. The next day a boiling ley is prepared from the ashes of the burnt vine cuttings, to which are added some handfuls of lavender, rosemary or other aromatic herbs. A bunch is plunged three times in succession into this ley. If the berries are slightly cracked, the ley is strong enough, but if they are much cracked, it is too strong. When it is properly prepared it is allowed to cool and settle; it is then strained through a linen cloth and a second time placed over the fire. When it boils, each bunch is dipped into it three times in succession. They are then spread on the hurdles, which are exposed to the sun during the day and taken into the house at night. The raisins are commonly completely dried at the end of two or three days.

The Zante grapes undergo a different treatment. They are cut some days after they have attained their complete maturity. They are deposited on hurdles very close together, or on cloths placed in the full sun. When the berries preserving the pedicle begin to be detached from the main stalk, they are lightly beaten with little sticks, in order to hasten this result. They are then passed through a sieve in order to separate them from the stems, and lastly subjected to the action of a fan or winnowing machine, in order to remove the dust and rubbish.

www.ingramcontent.com/pod-product-compliance
Lightning Source LLC
LaVergne TN
LVHW052226200926
848481LV00001B/222
9781425539849